Richard Lydekker

Life and rock

A collection of zoological and geological essays

Richard Lydekker

Life and rock
A collection of zoological and geological essays

ISBN/EAN: 9783741119033

Manufactured in Europe, USA, Canada, Australia, Japa

Cover: Foto ©Klaus-Uwe Gerhardt /pixelio.de

Manufactured and distributed by brebook publishing software
(www.brebook.com)

Richard Lydekker

Life and rock

LIFE AND ROCK:

A COLLECTION OF ZOOLOGICAL AND GEOLOGICAL ESSAYS.

By R. LYDEKKER, B.A. Cantab., F.G.S., F.Z.S.,

AUTHOR OF "PHASES OF ANIMAL LIFE" AND "HORNS AND HOOFS";
JOINT-AUTHOR OF "THE STUDY OF MAMMALS" AND
"A MANUAL OF PALÆONTOLOGY"; EDITOR AND
CHIEF AUTHOR OF "THE ROYAL NATURAL
HISTORY," ETC., ETC.

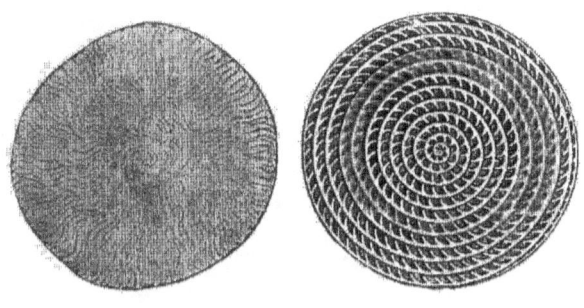

NUMMULITE.

𝔏𝔬𝔫𝔡𝔬𝔫:

THE UNIVERSAL PRESS, 326, HIGH HOLBORN, W.C.

1894.

PREFACE.

In presenting to the public the following series of essays, which have already appeared in serial form, the Author has but few words to say by way of preface.

It will be seen from the chapters themselves that the greater portion of the work relates to Zoology—and especially its palæontological branch—while only a small section is devoted to Geology proper.

While some of the zoological chapters treat of the natural history of particular groups or species of animals, others are more especially devoted to some of those problems connected with the evolution, development, and mutual relationships of animals which at the present day are attracting so much attention on the part of students of the science.

Although necessarily somewhat technical in places, the Author hopes that the popular style, which he has endeavoured to preserve throughout the work, will render it acceptable to those lovers of Nature who are too often repelled by the number of hard words they meet with in the books professedly written for their edification.

HARPENDEN,
    March, 1894.

CONTENTS.

LIST OF ILLUSTRATIONS.

— —◆———

LIFE AND ROCK:

A COLLECTION OF ZOOLOGICAL AND GEOLOGICAL ESSAYS

CHAPTER I.

ELEPHANTS, RECENT AND EXTINCT.

ASSUREDLY of all the mammals now inhabiting this earth elephants are those most justly entitled to the epithet "antediluvian," since they remind us, far more vividly than any of their modern contemporaries, of the gigantic extinct mammals of various kinds which flourished in that latest epoch of geological history when man was but a comparatively new comer. A long acquaintance has, indeed, made us so familiar with the appearance of elephants that we are too apt to forget what altogether strange and uncouth creatures they really are. If, however, they had happened to be included among those animals which disappeared from the face of the earth before the historic period, and were known to us solely by their skeletons, there can be no doubt that they would be regarded as among the most remarkable of mammals. Moreover, if elephants were only known to us by their skeletons it would be more than doubtful if we should ever have attained a correct idea of their true form; since although the conformation of their jaws and teeth would clearly indicate that they must have had some very peculiar method of feeding, yet it would have required a very bold, not to say a very imaginative man to have conceived the idea that these creatures were furnished with that unique organ which we term the trunk or proboscis.

At the present day, it need scarcely be said, there are but two living species of elephants, differing remarkably from one another not only in external characters, but also, as we shall notice later on, in the structure of their

B

teeth; these two species being respectively confined to
Africa, and to India and the adjacent regions.   These two
kinds of elephants are, however, merely the last survivors
of a vast host of extinct forms, some of which were closely
related to their living cousins; while others differed so
markedly in the structure of their teeth as to have
received the distinctive appellation of mastodons, although

Fig. 1.—Young African Elephant.

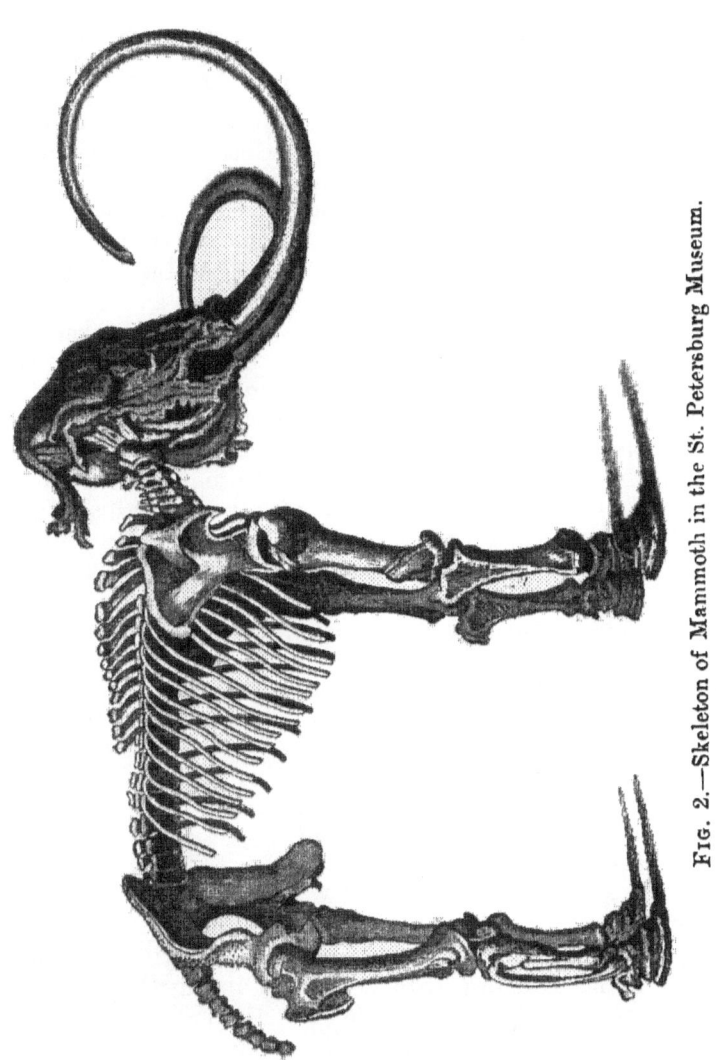

FIG. 2.—Skeleton of Mammoth in the St. Petersburg Museum.

they are really nothing but very generalized elephants. These so-called mastodons carry us backwards to the middle portion of that division of the Tertiary period of

B 2

the earth's history known as the Miocene ; but when we
have reached to that stage all below is dark as regards the
elephantine pedigree.   And it is, indeed, one of the most
remarkable circumstances in palæontology that although
we know that elephants belong to the great group of
Hoofed or Ungulate Mammals, of which they form a
well-marked division, yet we have practically no sort of
knowledge of the many extinct forms which we presume
must have connected them with Ungulates of a more
ordinary type.

Although the trunk and tusks of elephants form their
most striking external features, yet it is not to these that
the naturalist looks at first when inquiring into the true
affinities and general structure of these animals, since
these come under the category of specialized and acquired
structures, which tell but little of an animal's past history;
he looks rather to the structure of the internal skeleton,
which is always of especial value, as being that part of the
organism which is usually alone preserved in a fossil state.
Let us then first turn our attention to the skeleton of these
animals, of which we may see examples in our larger
museums. The most remarkable feature noticeable in such a
skeleton is that the various long-bones of the limbs are
placed almost directly one above another, so as to form
nearly vertical columns of support for the body ; whereas in
ordinary Ungulates, such as a horse or an ox, these bones
are set very obliquely to one another.   Moreover, as a
similar vertical position of the limb-bones occurs in several
old extinct Ungulates which are known to be of extremely
primitive organization, we may take it that an elephant's
limbs are likewise of a primitive type.   We have, however,
further evidence in confirmation of this primitive structure.
Thus elephants differ from all other living Ungulates in
having five complete toes to all their feet (Fig. 3).
Moreover, whereas in other living Ungulates (except the
little hyrax) the bones of the wrist are not situated in
vertical rows immediately over the metacarpal bones of
the foot, but, on the contrary, cross and overlap one
another, in elephants they have the former relation, with
the single exception that the bone marked l overlaps the

one lettered *td* to a certain extent. This difference may be illustrated by saying that if we were to take a hatchet and chop vertically up-wards between the third and fourth toes of an elephant's foot there would be nothing to resist the passage of the blade till it reached the bones of the leg, whereas in all other living Ungulates, except the hyrax—the pig for instance—the blade could not pass through the wrist without cleaving solid bone. Again, where-as ordinary Ungulates walk solely on the tips of their toes, and are thus termed digitigrade, while the bones of the toes them-selves are more or less elongated, elephants walk

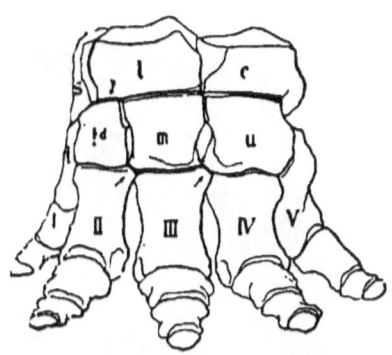

FIG. 3.—Bones of the Left Fore Foot of an Elephant, one-eighth natural size. The lettered bones are those of the wrist or carpus, and the numbered ones the metacarpals, below which are the bones of the toes. (After Osborn.)

on the soles of their feet in the so-called plantigrade fashion, and have very short toe-bones. Now, since all the large extinct Ungulates of the lower Eocene, or earliest Tertiary period, also have five-toed feet, very similar to, but still shorter and of even simpler structure than those of elephants, there can be no doubt as to the extremely primitive plan on which the entire limbs of the latter are constructed. As regards, therefore, its limbs and feet, an elephant may be said to be an essentially old-fashioned animal.

If, however, we turn to their teeth we shall find that elephants are very far indeed from being of a primitive or old-fashioned type; the truth being that they are, on the contrary, very peculiar and specialized in this respect. The first and most obvious peculiarity in regard to their dentition is to be found in their tusks, which correspond to one of the pairs of upper front teeth in man, and also to the single pair of such teeth in the Rodents (rats, hares, &c.). Moreover, these teeth, like the incisors of the Rodents, grow continuously throughout the life of the animal, owing to the circumstance that the pulp-cavity at their base always remains open, and has a permanent connection with the soft structures of the gum. In our own teeth, on the contrary, the pulp-cavity closes at a certain period, after which there is a total cessation of growth. These ever-growing tusks of the elephants are preceded in the young animal by a pair of small milk-tusks, with a closed pulp-cavity, which are shed at an early period of life. In both of the living species of elephant the tusks are confined to the

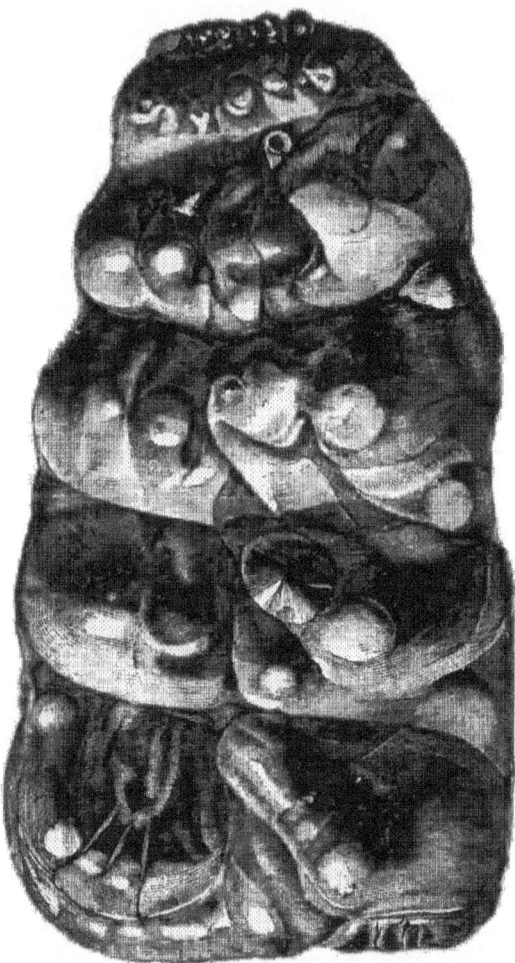

Fig. 4.—Last or Sixth Upper Molar Tooth of a Mastodon; half natural size.

upper jaw; but whereas they occur in both sexes in the African elephant, in the Indian species large permanent tusks are restricted as a rule to the male, and are not, indeed, invariably present in all individuals of that sex. In the female Indian elephant the tusks are usually very small, and soon shed. The primitive elephants, or mastodons, frequently, however, had tusks in both the upper and lower jaws; and since these did not generally attain the huge dimensions which they reach in many true elephants, it is evident that in this respect the mastodons departed less from the ordinary type of mammals, where the front teeth are not greatly larger than the hinder ones, and those of the upper and lower jaws correspond with one another in size and number. Before leaving the subject of tusks, it may be mentioned that the ivory of which they are composed differs from the so-called ivory of other teeth in a manner which renders it always easy to determine whether a reputed ivory article is genuine. This peculiarity consists in the circumstance that a transverse section of a tusk exhibits a series of fine, decussating, curved lines radiating from the centre to the circumference, and forming curvilinear lozenges at their intersections. This remarkable structure is in fact precisely similar to the "engine-turning" on the back of a watch; and in an ivory knife-handle it should be distinctly visible at the butt.

We turn now to the grinding or molar teeth, which present far more remarkable peculiarities. So peculiar, indeed, and unique are the structure and mode of succession of the molar teeth of elephants, that they are often very imperfectly understood, even by those who have spent half their lives among these animals. For instance, we find the late veteran elephant-hunter, Sir S. Baker, in his "Wild Beasts and their Ways," making a statement in regard to elephants that "both the Indian and the African varieties have only four teeth," whereas, as a matter of fact, every elephant has, in the course of its life, six molars on each side of both the upper and lower jaws. Instead, however, of all these teeth being in use at one and the same time, as are those of a cow or a horse, in existing elephants there are never more than portions of two teeth in use at any one time, although, in the extinct

mastodons, portions of three may co-exist. In all
elephants, the more anterior of these six teeth are smaller
and of simpler structure than the hinder ones, and the

Fig. 5.—The Indian Elephant.

whole series is protruded in an arc of a circle from back
to front, the larger hind teeth being pushed up and
gradually coming into use as the small anterior ones are

worn away and finally discarded. The hinder ones of these teeth (Fig. 6) are so large that while the front portion is being worn away, the back is still bedded in the gum. In the earlier elephants, or mastodons, these molar teeth are composed of a series of relatively low and widely separated transverse ridges, more or less completely divided into inner and outer moieties, and with large open valleys between them. Moreover, in all the teeth, except the last, these ridges do not exceed four in number, although in the sixth tooth (Fig. 4) there may be as many as five or six. Such a molar tooth can be easily derived from the ordinary type of tooth presented by the molars of the pig, in which the crown carries four columns, severally placed at its corners. When a tooth like the one represented in Fig. 4 becomes worn down by use, the enamel coating each of the columns would be cut through, so as to expose a series of more or less trefoil-shaped surfaces of ivory, each surrounded by a ring of the hard enamel. And it will be obvious that a tooth thus constructed of alternations of substances of different degrees of hardness would act as an efficient millstone. Elephants do not appear, however, to have been by any means satisfied with this comparatively simple kind of tooth, for as we pass upwards in the geological scale we find that there has been a gradual increasing complexity in the structure of the molar teeth of these animals, this being due to a graduated increase in the height of their transverse ridges, accompanied by an increase in the number of the ridges themselves. There is, indeed, an almost perfect structural gradation, now known to exist between the mastodon tooth, represented in Fig. 4, to the teeth of true elephants shown in Fig. 6. Both the latter examples are in a somewhat worn condition, but it will be readily seen that the lozenge-shaped surfaces of ivory, surrounded by enamel, in the tooth of the African elephant, correspond to the transverse ridges of the mastodon's tooth. In the true elephants, however, the open valleys between these ridges (which have now assumed the form of tall, thin, and nearly parallel laminæ) have been completely filled up by a third constituent of the tooth, known as the cement. The grinding surface of the tooth of such an elephant consequently consists of a solid mass, made up of alternating vertical transverse layers of various substances, arranged

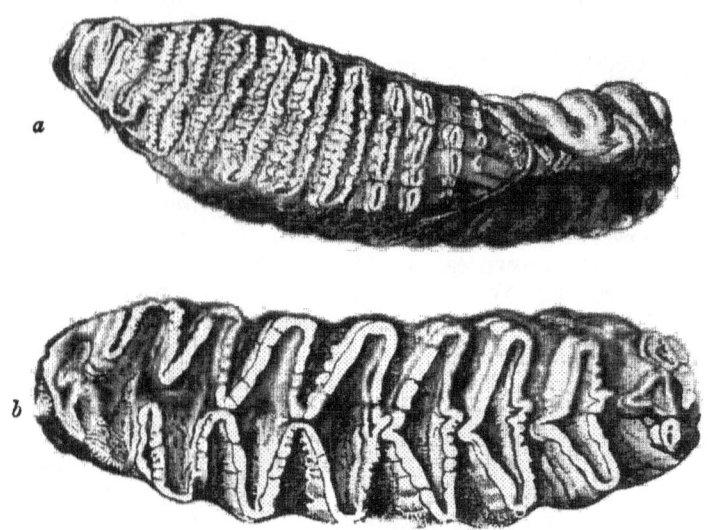

FIG. 6.--Molar Teeth of Indian (*a*) and African (*b*) Elephant. In *a* the anterior half is worn, and the remainder unworn. Much reduced.   (After Owen.)

in the order of cement, enamel, ivory, enamel, cement; and since each of these constituents differs in hardness, it will be obvious that the millstone-like apparatus is now of a far more efficient type than it was in the mastodon. Moreover, since the crowns of the molars of the true elephants are very much taller than are those of the mastodons, it is evident that they will require a longer period of time before they become worn away, and that they will therefore allow a longer life to their owner. Even, however, among true elephants there is a considerable amount of difference in regard to the number and narrowness of the component plates of their molar teeth, and it will be seen from Fig. 6 that in this respect the African elephant departs far less widely from the mastodon than does its Indian cousin.   Since the food of the latter consists to a large extent of boughs and twigs, while that of the former is composed more of juicy leaves, fruits, and roots, the necessity of a more complicated masticating apparatus in the one than in the other is apparent.

In modern elephants, the six molar teeth on either side of each jaw are all that are ever developed. This is, however, by no means the case with some of the earlier elephants, and with most of the mastodons. In these animals, when the second and third molars became worn out, they were succeeded vertically by much smaller teeth, from which we learn that the first three molars of the modern elephants really correspond to the milk-teeth of other mammals, which, as we all know, are succeeded vertically by some of the teeth of the permanent series. This succession of the teeth shows us, therefore, another point in which mastodons tend to connect modern elephants with ordinary Ungulates.

We might go further and enter upon the consideration of some of the structural peculiarities presented by the soft parts of elephants. Enough has, however, been stated to show that while these animals have preserved a very ancient type of structure in their limbs, they have acquired a very special modification in the structure and mode of succession of their teeth. And it is highly probable that it is owing to this particular specialization that elephants have survived to our own day, while all the other plantigrade and five-toed primitive Ungulates have completely passed away; while it is almost certain that it is this feature alone which has enabled them to attain the gigantic bulk which forms one of their most striking features. In regard to their evolution, perhaps no group shows more clearly than that of the elephants how exceedingly important is the study of fossils to elucidate the relations of existing animals. Had we only the two living species of elephants to deal with, we should never have had the least inkling of the manner in which they were related to other Ungulates, imperfect as our knowledge of the relationship still is. Moreover, from the distribution of these two species, it would have been naturally inferred that elephants were creatures suited solely to tropical or sub-tropical climates. The discovery of frozen carcases of the mammoth—a species closely related to the Indian elephant—in the Siberian "tundras" shows, however, that this animal was suited to dwell in at least comparatively cold regions, although it is probable that the climate of Siberia was formerly not so rigorous as at present. To withstand the cold of these northern regions, the mammoth was protected by a coat of long reddish hair, beneath which was a shorter covering partaking more of the nature of wool. Along the borders of the Arctic Ocean for hundreds of miles mammoth remains are met with in incredible quantities; and it is still one of the puzzles of geology to account adequately and satisfactorily for the manner in which these creatures perished, and how their bodies were buried beneath the frozen soil before decomposition had begun its work, for it is hardly possible to believe that they lived in a climate so rigorous that their bodies would have been frozen on the surface of the ground immediately after death.

Another rude shock to our common ideas of elephantine nature is afforded by the extinct elephants of Malta, which show us that gigantic size is not a necessary concomitant of the group; and that when the area in which a species dwelt was small, the size of the species itself was proportionately reduced. These little Maltese elephants were very closely allied to the living African species, but whereas "Jumbo" attained eleven feet in height, and wild specimens of the African elephant may be still larger, the smallest of the Maltese species was scarcely taller than a donkey. So small, indeed, are the bones and teeth of this species exhibited in the Natural History Museum, that it is sometimes difficult to convince people that they really belong to elephants at all.

As regards their distribution, elephants and mastodons formerly roamed over the whole world with the exception of Australia; true elephants ranging over the whole of the Northern Hemisphere, while mastodons extended as far south in the New World as the confines of Patagonia. It is in the north-east of India, Burma, and the islands of the Malayan region that the fossil elephants connecting the living species with the mastodons are alone found; and it is thus probable that from these regions the true elephants migrated westward into Europe and Africa, while the mammoth in later times crossed from Asia into Alaska by way of Behring Strait. That the mammoth, which ranged from the Arctic regions to the Alps and Pyrenees, was a contemporary of the primeval hunters of Europe is now a well-established fact, but it appears that throughout the Old World mastodons had utterly died out before the advent of man. In the New World, however, the continuity between the old and the new fauna was more fully sustained, the Missouri mastodon having survived well into the human period, so that we have in this survival a good instance of the vast changes that have taken place in the fauna of the globe within what we may metaphorically call the memory of man.

CHAPTER II.

TUSKS AND THEIR USES.

MANY mammals, such as the elephant, hippopotamus, and walrus, are furnished with one or two pairs of pointed conical or compressed teeth largely exceeding all the others in length, and to which the term "tusks" is usually applied.   Lions and wolves likewise have two pairs of somewhat similarly enlarged teeth in the fore part of their jaws ; and although these are proportionately smaller than in the animals above named, it will be obvious that in popular language it is difficult not to include them under the same general title.   Using, then, the term "tusks" for all such enlarged simple teeth, it may be of interest to note the groups in which these attain their greatest development, and also endeavour to learn something regarding their uses.   An especial interest attaches, indeed, to the subject, on account of the extreme beauty of many of these teeth, and also from the circumstance that it is these alone which yield the various descriptions of ivory.

In a great number of instances such tusks comprise a pair in both the upper and lower jaws, which are situated immediately behind the front, or incisor teeth, and, from their marked development in the dog tribe, are scientifically designated canine teeth ; the name "eye-teeth" being also not unfrequently applied to them in popular language.   Tusks of this sort are characterized by the circumstance that the lower one on each side bites in front of the upper one ; while the latter is always the first tooth situated in the true upper jaw-bone, the incisors, or front teeth, being implanted in a more anteriorly situated bone known as the pre-maxilla.   In any ordinary carnivorous mammal, such as a lion or a wolf, the upper tusks are considerably larger than the lower, although neither project beyond the edges of the muzzle when the jaws are closed. If we examine such tusks in a dried skull we shall find that their roots (which, as in all tusks,

Fig. 7.—Skeleton of Missouri Mastodon.

are simple) are completely closed ; thus indicating that their growth ceased at a certain period of life. Moreover, we may notice that the upper and lower pairs of tusks do not abrade against one another to any marked extent, and that consequently their summits are only subject to the ordinary wear and tear necessarily undergone during use. In many other mammals the form and structure of these tusks is, however, very different. Take, for instance, a wild boar, in which the upper tusks are short and curved upwards, while the larger and more slender lower pair abrade against their outer surfaces, and are thus worn to sharp, cutting edges. Obviously such tusks, if they were incapable of growth like those of the lion, would soon be worn away to mere stumps and become useless to their owners. To prevent this, the bases of the tusks remain permanently open (as shown in Fig. 10), and contain a soft pulp connected with the vascular structures of the jaw, in consequence of which the teeth continue to grow throughout life ; their rate of growth thus keeping pace with that of the abrasion to which they are subject. Tusks may accordingly be divided into two classes, which we may designate hollow and solid. The open tusks referred to above are prevented from attaining any very great length by the abrasion of the lower against the upper pair, but in other cases, as in those of the remarkable pig of Celebes known as the babirusa (Fig. 8), no such abrasion takes place, and both pairs then attain enormous dimensions, projecting in this particular instance far above the upper surface of the head, and the down-curving points of the upper pair sometimes even penetrating the skull. A babirusa may in fact be compared to a wild boar in which one pair of tusks having been broken, the other continues to grow without any abrasion by wear ; only that in these animals both pairs are thus developed. In the wild boar and its allies the tusks—more especially those of the lower jaw—are purely offensive and defensive weapons ; but it is hard indeed to imagine the use of those of the babirusa. Probably Mr. Wallace is right in regarding them as an ultra-development of organs originally useful, but which have now, from some reason or another, become of no functional advantage to their owner, and have thus, so to speak, run riot.

The babirusa presents us with an instance of a mammal in which both pairs of tusks have acquired their enormous development owing to the cessation of the mutual attrition between those of the upper and lower jaws, characterizing all its allies. On the other hand, in the mastodons and elephants we have examples where the development is normal, and either one or both pairs attain a vast size.

In both species of existing elephants, as in many of the mastodons (Fig. 7), only the upper pair of tusks is thus developed; but in other mastodons these organs were present in both jaws, the upper pair being, however, always much larger than the lower. In the pigs and babirusa the tusks, of which the upper pair are implanted in the true jaw-bone, correspond severally with those of

Fig. 8.—Fore part of Skull and Tusks of Babirusa.

the lion and the wolf, and are accordingly reckoned as canines. This, however, is not the case with those of the elephants and mastodons, which grow in great part from the pre-maxillary bone, and thus correspond with one pair of the front or incisor teeth of other mammals. Consequently, much as they resemble them in general appearance, the tusks of an elephant are not homologous with those of a pig or a babirusa. As we have entered

c

in some detail into the structure of the tusks of the elephants in the previous chapter, it will be unnecessary to recapitulate the facts here; although we may mention that in modern elephants these organs consist wholly of ivory, without any investing coat of enamel. The tusks of these animals, which belong to the hollow type, are the largest developments of dental structure to be met with in the whole animal kingdom. To a great extent they are weapons of defence and offence, but in the African species, where they are common to both sexes, they are also largely used in grubbing up roots and overturning trees; while in an extinct species from India their length is so great that they must have quite ceased to be useful, and were probably an actual encumbrance. Another allied extinct animal, known as the dinothere, is unique in having a large pair of downwardly-bent tusks in the lower jaw and none in the upper; the use of these being very difficult to conjecture.

A still more remarkable condition obtains in the hippopotamus, in which not only are the canine teeth developed into an enormous pair of hollow, ever-growing curved tusks in each jaw, but the central pair of lower front or incisor teeth are so enlarged as likewise to merit the title of tusks. These incisor-tusks—which thus correspond to the lower pair of the four-tusked mastodons—are likewise of permanent growth, and project forwards from the front of the jaw in the form of two elongated cones. In thus possessing three pairs of tusks the hippopotamus is quite peculiar among animals. Largely employed for tearing up the grasses, on which these monsters feed, the tusks of the hippopotami are also most effectual offensive weapons.

In the land carnivores, of which more anon, the tusks are always of the closed type; but in their aquatic ally, the walrus, we again meet with a huge pair of ever-growing tusks directed downward from the upper jaw. On comparing the head of a walrus with that of an elephant, most persons would say at once that the tusks of the two animals were homologous. In this, however, they would be wrong, since, as we learn from the condition in the young animal, those of the walrus are true canines, whereas, as we have seen, the tusks of the elephant are incisors. We have here, therefore, a well-marked instance of the parallel development of severally dissimilar structures to attain a marked general similarity. The tusks of the walrus, which in old animals attain a great length, are mainly employed in digging up molluscs and crustaceans from the sand and shingle, and also, it is said, to enable their owners to clamber up on the ice.

Although the whalebone-whales are entirely deficient in teeth, and many of the dolphins and their allies have these organs but poorly developed, there are two cetaceans which exhibit a most remarkable development of tusks. The first of these creatures is the well-known narwhal, of the Arctic seas, in which, as a rule, there is one huge spirally-twisted cylindrical tusk projecting from the left side of the upper jaw of the male, which continues to grow throughout life. Whether this solitary tusk is a canine or an incisor, is not very easy to determine ; but it is

Fig. 9.—The Narwhal. (After True.)

remarkable that its fellow of the opposite side generally remains concealed in the jaw-bone, like the kernel of a nut in its shell, while in the female both teeth are thus rudimentary. Occasionally, however, male narwhals are met with in which both the right and left tusks are developed ; and it is somewhat curious that in such cases the direction of the spiral in the two tusks is the same, instead of being, as in the horns of antelopes, opposite. Although narwhals have never been known to charge and pierce ships with their tusks, after the manner of sword-fish, it is still uncertain whether these formidable weapons (which may attain a length of from eight to nine feet) are normally

c 2

used for purposes of attack, or for procuring food. Doubtless, however, the narwhal's tusk is of some use to its owner; but in another cetacean, known as Layard's mesoplodon, the tusks appear not only useless, but actually harmful. In the whale in question, which is a rare species from the southern seas, there is but one large strap-like tusk on each side of the middle of the lower jaw, both of these curving upwards and inwards over the snout, so as actually to prevent the mouth from opening to its full extent. The only possible use we could suggest of such a structure would be to prevent the creature dislocating its jaw by yawning; but as other animals manage to get on without such an arrangement, this is scarcely likely to be a solution of the problem. It is more probable, indeed, that we have here to do with another instance of ultra, or monstrous development.

Other examples of hollow or permanently-growing tusks

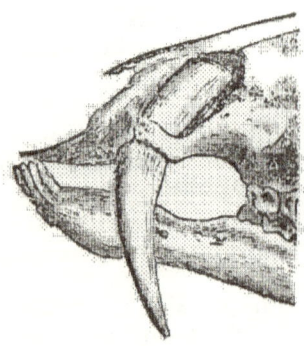

occur among the hoofed mammals, other than the pigs, in all of which these teeth are found only in the upper jaw, and are developed chiefly or solely in the males. Among recent forms these tusks attain their greatest development in the little musk-deer of the Himalaya, where they are frequently over three inches in length, and project considerably below the lower jaw. In form they are sabre-like, and recall the upper tusks of the feline carnivores, only being more slender, and grow-ing permanently. Similar but

Fig. 10.—Extremity of the Skull of a young Chinese Water-Deer, with the base of the Tusk exposed. (After Sir V. Brooke.)

smaller tusks are met with in the Chinese water-deer (Fig. 10), in the Indian muntjac, and the little deer-like animals known as chevrotains. The latter belonging to a totally distinct group from the others, it is evident that these scimitar-like tusks have been inde-pendently acquired in the two groups; while it is quite probable that those of the musk-deer and Chinese water-deer are likewise of separate origin. With the exception of the muntjac, in which they are very small, all these deer-like animals are devoid of antlers; and it is thus evident that their tusks have been developed in lieu of those weapons. It is true, indeed, that the males of some of the antlered deer have small tusks, but these are of no use for offensive purposes, and are evidently organs in process of degeneration. Moreover, in the hollow-horned ruminants, such as oxen and antelopes, where the horns are permanent and generally present in both sexes, not a vestige of tusks remains.

Although the Asiatic rhinoceroses have procumbent tusks of considerable size in the lower jaw, none of the odd-toed hoofed mammals, such as horses and tapirs, have upper tusks of any size. In past times there were, however, in North America a group of somewhat allied creatures known as uintatheres, in which an enormous pair of upper tusks, somewhat like those of the musk-deer, was developed (Fig. 11). Like those of the latter, these tusks grew permanently, and they were also protected by a descending flange of the lower jaw, which was often deeper than in the figured example. As these animals had front teeth only in the lower jaw, we have here, therefore, another curious instance of the parallel development of similar conditions in totally unconnected groups. What use these creatures could have made of their tusks is, however, not very clear, as in the living condition they could have projected but little below the lower jaw, while they were too long to have been effectual when the mouth was open. These uintatheres were also noteworthy on account of having a number of bony projections on the top of the skull, which in life may have been sheathed in horn; and it is not a little remarkable that another extinct creature—*Protoceras* (Fig. 19)—belonging to the even-toed group of hoofed mammals had a skull with very similar bony projections and also similar upper tusks.

Our last instance of animals furnished with permanently-growing upper tusks is afforded by the reptilian class, in which the extinct South African creatures

described many years ago by Sir R. Owen, under the
name of *Dicynodon*, are thus armed. In these reptiles,
some of which attained gigantic dimensions, the jaws were
mainly sheathed in horn like those of a turtle; the single

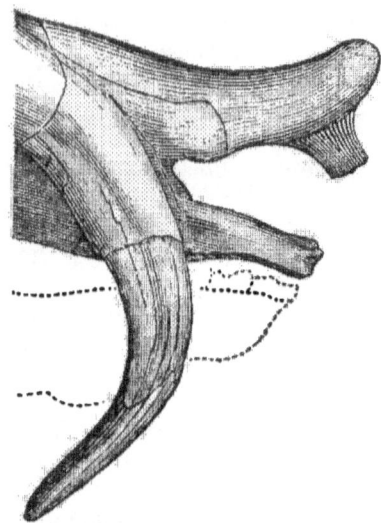

pair of tusks curving downwards and forwards from
near the middle of the upper
one. Possibly these tusks
were capable of being em-
ployed when the mouth was
open like those of lions or
tigers; but otherwise it is
difficult to see their use.
Certain allied reptiles from
the aforesaid formations,
known as anomodonts,
had a full series of teeth,
with a large pair of tusks
in the upper jaw, which
may or may not have been
rooted, but are exceedingly
like those of certain carnivores. As these reptiles
are not the direct ancestors
of mammals, we have
thus evidence of the acquisition of tusks in two distinct
classes.

FIG. 11.—Extremity of the Skull
and Tusk of a Uintathere.

As regards solid tusks, or those in which the lower end
is closed at a certain period coincident with the cessation
of growth, almost the only mammals, save the rhinoceroses
(where, as we have seen, there may be a forwardly-directed
pair in the lower jaw) and some of the marsupials, in
which they attain any marked development are the carnivores. Among these, the maximum size of tusks at the
present day is attained in the larger felines, such as the
lion, tiger, and leopard. In these animals the tusks are,
however, never so much elongated as to bar the front of
the open mouth; while very frequently their tearing
power is increased by the hinder cutting-edge being
finely serrated. Their terrible effect in tearing and

rending the animals upon which these fearsome carnivores prey is too well known to need further mention. During the Tertiary period there existed certain carnivores nearly allied to the modern cats, but exhibiting a much greater development of the upper pair of tusks, with a corresponding reduction of those of the lower jaw. In these creatures, which are known as machærodonts, or sabretoothed tigers, the upper tusks were greatly compressed and flattened, with serrations on one or both cutting-edges; their length in a species of the approximate size of a tiger being upwards of seven inches. As in the uintatheres, the anterior end of the lower jaw had a descending flange to protect the end of the tusk. Manifestly, such enormous weapons would completely bar the sides of the open mouth, and consequently they could not be used in the manner of the tusks of a lion or tiger. It is still more difficult to imagine that these animals could have struck with their tusks projecting from below the closed mouth; and it would consequently seem that in this instance also the tusks attained a development which was harmful rather than advantageous—this conclusion being confirmed by the fact, for what it is worth, that the sabre-toothed tigers have become totally extinct, while their less specialized allies continue to flourish.

Finally, as regards tusks in general, it appears that while in the land carnivores these are always canines, are present in both jaws, and have closed roots, in the other orders of mammals their development is apparently somewhat capricious, while they are very frequently present in only one jaw—almost invariably the upper—and continue to grow permanently. Moreover, they are almost as frequently incisors as they are canines; so that tusks apparently similar may be in nowise homologous with one another. Never developed to any size in animals with large cranial appendages in the form of antlers or horns, tusks are frequently wanting in those lacking the latter. Primitively their use was undoubtedly as weapons of attack and defence, or to aid in procuring vegetable food; but in many cases they have subsequently undergone a (frequently sexual) development beyond the needs of such purposes, and are thus in this respect analogous to the antlers of many stags. In other instances, however—and this in all the groups in which they occur—they have undergone a still further semi-monstrous development, rendering them if not actually harmful, probably in some cases inconvenient to their owners. Lastly, the independent acquisition in closely allied or widely separated groups of mammals of tusks of very similar structure and appearance, shows how little reliance is to be placed on external characters as indicative of relationship.

CHAPTER III.

MOLES AND THEIR LIKE.

It is probably well known to most of our readers that in the evolution of organized nature two great factors have constantly been working against each other—the one being the adherence to a particular type of structure, while the other is the adaptation to a special mode of life. The usual resultant of these two forces has been that, in any assemblage of animals specially adapted for a certain peculiar kind of existence, while internally its different members have preserved their essential structural peculiarities more or less intact, externally they have become so much like one another that it often requires the aid of the professed zoologist to point out their essential distinctness. Perhaps in no case is this adaptive similarity in external characters better displayed than among certain of the smaller mammals which have taken to a more or less completely subterranean burrowing existence, of which the common mole is the best known example. In the British Islands we have, indeed, only this one creature which has adopted this particular mode of life; and it is to this animal alone that the name "mole" properly belongs. Other parts of the world possess, however, several more or less closely allied animals to which the same name must clearly be also applied. If, however, we happen to have friends from the Cape, we may hear them applying the name "moles" to certain burrowing mammals from that district, which upon examination would be found to differ essentially in structure both from the ordinary moles and from one another. Then, again, if we were to travel in Afghanistan or some of the neighbouring regions, we should meet with another mole-like burrowing animal to which we should likewise feel disposed to apply the same name, although it has not the most remote kinship with our English mole. Finally, the deserts of central South Australia are the dwelling-place of the "marsupial mole," which, although mole-like in general form, differs from all the animals yet mentioned in belonging to the marsupial order.

We thus arrive at the conclusion that in the popular sense the term "mole" now serves to indicate a number of widely different animals, whose sole or chief bond of union is to be found in their adaptation to a similar mode of life, and their consequent assumption of a more or less similar outward form. Hence, in order to avoid confusion, it will be necessary to prefix the epithet "true" to those species which belong to the same family group as the "little gentleman in black velvet," while the remainder must be designated by other distinctive epithets. It might have been thought that such an expanded application of the name "mole" was restricted to popular language. This, however, is not the case, as naturalists have found it convenient to adopt the names "sand-mole," "golden mole," "marsupial mole," etc., as the distinctive titles of different members of this purely artificial assemblage of animals; and the reader will accordingly understand that when we speak of "moles and their like," we merely refer to a similarity in habits, and a more or less marked external resemblance between the animals under consideration.

The general bodily form of the common mole is so thoroughly well known and familiar, that the term "mole-like" has been introduced into zoological, if not into popular, literature as a definite descriptive epithet. Since it is perfectly obvious that this peculiar form is the one best adapted for the needs of the creature's subterranean existence, no explanation is necessary why most of the other members of the assemblage have conformed more or less closely to the same type. We may especially notice the flat, tapering, and sharp-nosed head, passing backwards without any distinctly defined neck into the long and cylindrical body; the comparative shortness of the limbs, and the immense strength of the front pair, which are placed close to the head, and have their feet expanded into broad, shovel-like organs. We shall also not fail to observe the absence of any external conchs to the ears, and the rudimentary condition of the deeply-buried eyes. A long tail would also be useless to a burrowing animal, and we accordingly find this appendage reduced to very small dimensions ; while the close velvety hair is most admirably adapted to prevent any adhesion of earthy particles during the mole's subterranean journeys. Equally well-marked adaptive peculiarities would also present themselves were we to undertake an examination of the mole's skeleton. While the majority of the assemblage conform more or less closely to the true moles in appearance, there are others in which such resemblance is but slightly marked, if apparent at all, from which we may probably infer some minor differences in their mode of life.

Proceeding to the consideration of the different groups of mole-like animals, it will be convenient to divide the mixed assemblage into insectivorous moles, rodent moles, and marsupial moles. The term "insectivorous moles," it may be premised, does not primarily indicate carnivorous habits in the species thus designated, but merely refers to the fact that they are members of the order Insectivora. It would of course be impossible, not to say out of place, to attempt a definition of that order ; but it may be mentioned that it includes small mammals, like shrews, moles, and hedgehogs, which differ from the rodents in not possessing a pair of chisel-like teeth in the front of the jaws, and also have their molar teeth surmounted by a number of small sharp cusps.

The insectivorous moles include not only our common mole (*Talpa europæa*), but also many other species belonging to the same genus, as well as certain others which are referred to distinct genera, all of which, for our present purpose, may be collectively spoken of as true moles. Of these, two genera (*Talpa* and *Scaptonyx*) inhabit Europe and Asia, while the other three are North American ; Africa having no representative of the group. It may be well to mention that all the true moles have very broad naked hands, each furnished with five toes carrying long flattened nails, in addition to which there is a sickle-like

extra bone internally to the thumb.   In burrowing, most
of them throw up the well-known molehills at certain
intervals from the tunnels driven in search of worms—
their chief food.

In addition to these true moles, North America also
possesses certain other species known as shrew-moles,
which, while belonging to the same family (*Talpidæ*), are
distinguished by the absence of the sickle-like bone in the
hand and the less expanded form of the bones of the upper
part of the fore-limb.   They are thus clearly seen to be
less specialized creatures than our own mole, to which
they closely approximate in general appearance.

Although belonging to the insectivorous order, the mole-
like creature represented in the accompanying figure
indicates a totally different family group.   If we were to
examine the upper molar teeth of a common mole, we
should find that they had broad crowns, carrying cusps
arranged somewhat after the manner of the letter **W**.   On
the other hand, in the Cape golden mole (as the animal

Fɪɢ. 12.—The Cape Golden Mole.

here represented is termed) the corresponding teeth have
triangular crowns carrying three cusps arranged in a **V**.

Moreover, if we look at the fore-limb, we find instead of the five-fingered hand of the mole that there are but four digits, of which the lateral pair are small, while the two middle ones are enlarged and furnished with triangular claws of great power. As in the true moles, all external traces of ears and eyes are concealed by the fur ; this latter, it may be added, having a peculiar golden-green metallic lustre, from which the name of the animal is derived. The golden moles, of which there are several species, are much smaller than our English mole, and are widely distributed in South Africa ; in which continent they, in conjunction with the under-mentioned sand-mole and its allies, take the place occupied in the northern hemisphere by the true moles. In tunnelling, the golden moles come so close to the surface as to leave a ridge marking their course. The true moles and the golden moles afford us, therefore, an instance of two entirely distinct groups belonging to the same order having assumed a perfectly similar mode of life, and, consequently, having acquired a superficial general similarity in external appearance.

With the rodent-moles, of which there is likewise more than a single group, we come to animals of a totally different order, which have assumed a mole-like form and habits, and are popularly confounded with the true moles. In common with the other members of the order Rodentia, all these rodent-moles are characterized by the presence of a pair of powerful chisel-like incisor teeth in the front of each jaw, while their molars have broad and flattened crowns adapted for grinding. Moreover, instead of driving their tunnels in search of worms, these rodent-moles burrow for roots and bulbs. All of them have very small or rudimentary ears and eyes, large and powerful claws, and short tails.

One of the best known of these rodent-moles is the great mole-rat (*Spalax*), ranging from south-eastern Europe to Persia and Egypt, in which the eyes are completely covered with skin ; allied to which are the bamboo-rats (*Rhizomys*) of north-eastern Africa and Asia, distinguished by having minute uncovered eyes and small naked ear-

conchs, and thus departing more widely from the mole type.
In the sandy soil of Egypt the mole-rat constructs tunnels
of great length in search of bulbs. In South Africa these
forms are replaced by the huge sand-mole (*Bathyergus*),
which attains a length of about ten inches; and also by
certain smaller animals known as *Georychus* and *Myoscalops*,
differing from the former by the absence of grooves in
their incisor teeth. The sand-mole is commonly met with
in the flats near the shore, while the smaller forms
generally frequent land at a higher elevation. Sometimes,
however, both are found together, and the country is then
covered in all directions with hillocks precisely resembling
those made by our English mole. Although the sand-mole
has uncovered eyes, these are not bigger than the heads
of large pins, and can have but little visual power. Still,
however, their presence serves to indicate that these
animals have not become so completely adapted to a
subterranean life as is the common mole, and this is
confirmed by the fact that if their burrows are opened,
the sand-moles after a few minutes usually protrude their
noses from the aperture with a view to discover the cause
of the disturbance, whereas an ordinary mole would under
similar circumstances remain below.

All the foregoing belong to one family of rodents; but
in addition to these certain members of the vole group
(a sub-division of the *Muridæ*) have also taken to a subterranean burrowing life, with the assumption of a mole-like
bodily form. These may be termed mole-voles, and range
from Russia to central and northern Asia, where they are
represented by the two genera *Ellobius* and *Siphneus*. They
all have mole-like heads and bodies, short limbs and tails,
rudimental external ears, very minute eyes, and powerful
fore-paws. In the Russian mole-vole (*Ellobius*) and the
allied Quetta mole from Afghanistan the claws of the front
paws are short; but, as shown in our figure, they become
greatly elongated in the members of the genus *Siphneus*.
All of them agree with the ordinary voles in the peculiar
structure of their molar teeth, which consist of a number
of triangular prisms placed edge to edge; and all are
described as driving subterranean tunnels and throwing

out at intervals heaps of earth precisely after the fashion of the common mole.

The foregoing are the only rodents which have assumed a more or less distinctly marked mole-like external form while retaining all the characteristic structural features of

FIG. 13.—The Long-clawed Mole-Vole.

the order to which they belong. There are, however, two other members of the same great order which, while having acquired mole-like habits, have not assumed a distinctly mole-like form. One of these is the tuco-tuco (*Ctenomys*) of South America, belonging to the same great family as the capivara and the coipu. This animal is rather smaller than a rat, with a relatively shorter tail, pale grey fur, and red incisor teeth. Its general form is also not unlike that of a rat, the limbs being of fair length, and the front paws not markedly enlarged, while the eyes are of considerable size. The external conchs of the ears have, however, been greatly diminished in size. The tuco-tuco derives its name from its voice, which resounds day and night from its subterranean dwellings, and is compared by Mr. W. H. Hudson to the blows of a hammer on an anvil.

It frequents loose and sandy soil, although occasionally found in moist heavy mould, through which it pierces its way as readily as the mole   Darwin, who states that the tuco-tuco is even more subterranean in its habits than the mole, was told by the Argentines that blind examples were often captured.   This, however, is not the experience of Mr. Hudson, who lays stress on the relatively large size of the creature's eyes.   From the soft nature of the soil in which it tunnels, it is not difficult to understand why it has been unnecessary for the tuco-tuco to assume a mole-like bodily form; but the reason for the retention of fully-developed eyes—which we should have thought exceedingly prone to injury—is hard indeed to divine.

The other rodents with mole-like habits are two tiny little creatures from the sandy districts of Somaliland, locally known by the name of farumfer, and scientifically as *Heterocephalus*.   They are about the size of a mouse, with large heads, moderately long tails, long powerful fore-feet, no external ear-conchs, minute eyes, and the whole skin naked, save for a few sparse bristly hairs. About as ugly a creature as can well be conceived, the farumfer, if clothed with a thick coat of fur, would be not very unlike a rather long-tailed, long-limbed, and narrow-handed mole.   For tunnelling beneath the hot sand of the Somali desert the naked skin of the farumfer is most admirably adapted; and as the creature is allied to the South African *Georychus*, it may be regarded as the member of this group most specially adapted for a subterranean existence.   Mr. E. L. Phillips, who was the first to observe these curious rodents in the living state, writes that they threw up in certain districts groups of elevations in the sand which may be compared to miniature volcanic craters. When the animals are at work, the loose sand from their tunnels is brought to the bottom of the crater and sent with considerable force into the air with a succession of rapid jerks, the rodents themselves remaining concealed in the shelter of their burrows, from which they appear never to venture forth.

The extensive assemblage of pouched or marsupial mammals contains groups corresponding to several of those of the placental or higher mammals. Thus, for instance, the kangaroos in Australia play the same *rôle* as the deer and ruminants in other parts of the world, while the Tasmanian wolf takes the place of the ordinary wolf, the wombats act the part of the marmots, the phalangers of squirrels, and the bandicoots of the civets and weasels. Till recently, it was thought that the place of the moles (whether insectivorous or rodent) was unoccupied in the Antipodes, and that no marsupial had adapted itself to a tunnelling subterranean existence. Within the last few years it has, however, been discovered that the sandy deserts of south Central Australia are inhabited by a small burrowing creature belonging to the pouched group, which has been fitly termed the marsupial mole (*Notoryctes*);

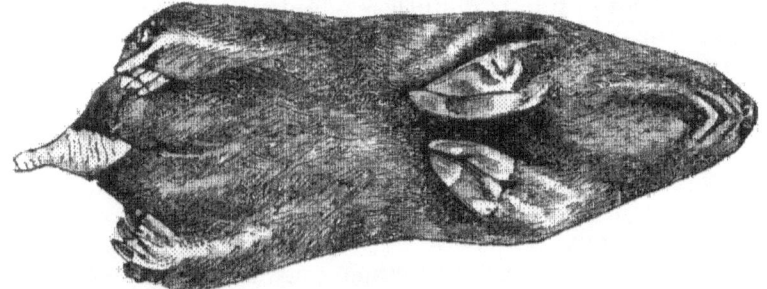

FIG. 14.—Under-surface of the Marsupial Mole, two-thirds natural size. (After Stirling.)

and it is not a little remarkable that in general appearance this tiny animal is even more mole-like than are some of the above-mentioned burrowing rodents, thus showing how all-powerful is adaptation to environment, and of how little import is internal structure in modifying the external form of an animal. The general mole-like appearance of the Australian burrower will be apparent from the accompanying figure; the most striking mole-like features being the elongated and depressed body passing imperceptibly into the head, the absence of external ear-conchs, the rudimentary pin-like eyes, the small tail, and the short limbs, of which the front pair are armed with claws of

D

great power. The marsupial mole stands alone, however, in having the front of the muzzle protected by a leathery shield; while its short and blunt tail is also covered with a peculiar naked leathery skin. In the fore limbs the structure of the feet recalls the golden moles rather than the true moles, the third and fourth toes being greatly enlarged at the expense of the others, and furnished with huge triangular claws of enormous digging power. In its pale, sandy-coloured hair, with a more or less golden tinge, the marsupial mole departs widely from our sable European friend; but it must be remembered that the difference in this respect is really not so great as it at first sight appears, seeing that cream-coloured varieties of the common mole are far from rare. In the Australian form the pale coloration is doubtless adapted to harmonize with the natural surroundings of its native desert, since it has been ascertained that the creature makes its appearance from time to time above ground. That the resemblance of the marsupial to the golden mole in the structure of the fore-paws is a purely adaptive one, there can be no reasonable doubt; but whether the identical structural conformation of the molar teeth of the two animals indicates any real genetic affinity, or is merely inherited from an old ancestral type which may have been common to many groups, is far less easy to answer.

The marsupial mole, or ur-quamata, as it is termed by the aborigines, inhabits a very limited area lying about a thousand miles to the interior of Adelaide; and even there appears to be of extreme rarity. According to observations supplied to its describer, Dr. E. C. Stirling, the creature is generally found buried in the sand under tussocks of the so-called porcupine-grass (*Triodia*), and its food appears to consist of insects and larvæ. The animals appear only to move about during warm, moist weather; and as they are extremely susceptible to cold, it is probable that they lie in a more or less torpid condition during the winter months, when the surface of the ground is often white with frost. When on the move, the marsupial mole is said to enter the sand obliquely, and to travel for a few feet or yards beneath the surface, when it emerges, and after traversing a short distance above ground, once more descends. As it seldom tunnels at a depth of more than two or three inches below the surface, its course may often be detected by a slight cracking or movement of the surface as the tunnel proceeds. Both in this respect, and in the nature of its food, the ur-quamata therefore exhibits a further analogy with the golden mole. In burrowing, the leathery shield of the head is said to be brought into play as a borer in the soft sand.

As regards the advantages obtained by those mammals which have taken to a burrowing subterranean existence, it will be manifest that these are twofold. In the first place, the creatures are secure from all foes, except those which, like the weasel and the snake, are able to follow them into their underground labyrinths ; while, secondly, they tap a food-supply (whether animal or vegetable) inaccessible to most other animals. In the case of the mole at least, whose habitations are generally made in comparatively hard ground, the life must be an incessant round of labour ; and to our thinking, at any rate, the existence of all these burrowing creatures must be so dull and monotonous as to leave no question as to whether it is worth living.

CHAPTER IV

SPINY ANIMALS.

In the preceding chapter it was shown how the adaptation to the necessities of a particular mode of life has produced a marked general external resemblance in certain burrowing mammals belonging to several more or less completely distinct groups. We now propose to point out the resemblances existing between certain other members of the same class of animals, owing to the assumption of a protective coat of spines. Although this resemblance is in some instances not so striking as among the creatures noticed in that chapter, yet it is quite sufficient to have obscured in popular estimation the real affinities of some of the spine-bearing mammals, as it is by no means uncommon to hear the hedgehog spoken of as the "British porcupine," while certain Madagascar spiny mammals are frequently alluded to as hedgehogs, and the Australian echidna is commonly alluded to as a porcupine. Moreover, the names "sea-urchins" and "sea-hedgehogs," applied to animals belonging to totally different classes, shows the important estimation held by spines in popular zoology. It is almost superfluous to add that the acquisition of the coat of spines in all the mammals here alluded to is solely for the purpose of protection : and how sufficient is this protection in most cases, is evident to all who have seen how the hedgehog, when rolled up, sets most dogs at defiance. Still, however, this panoply is by no means invariably proof against all attacks, as it appears to be well ascertained that leopards and pumas will kill and eat porcupines without the slightest hesitation, and with a total disregard of their formidable spines, which may be found sticking in all parts of the bodies of the devourers. As we found to be the case with the mole-like mammals, all the spiny mammals belong to the lower orders of the class, their

several representatives being distributed among the insectivores, rodents, and egg-laying groups, and the
majority pertaining to the two former of these. In fact,
in this respect an exact parallelism may be drawn between
the mole-like and the spiny mammals, each assemblage
having several representatives among the insectivores and
rodents, while the former has a solitary marsupial type,
and the latter two members among the egg-laying
mammals. Some of these spiny mammals, such as the
true porcupines and the echidnas, are burrowing creatures,

FIG. 15.- The Common Porcupine.*

and thus have a double means of defence against their
enemies; others, however, like the hedgehog, rely on their
power of rolling themselves up into a ball, and thus presenting a *chevaux-de-frise* on all sides. Some again, like
the tree-porcupines, are more or less completely arboreal
in their habits; and the whole of them, like the mole-like
mammals, indicate how urgent has been the need for the
lowly-organized rodents, insectivores, and egg-laying

* We are indebted to Messrs. F. Warne and Co. for the loan of
the three figures illustrating this chapter.

mammals to acquire some special means of protection in order to be able to hold their own among the higher forms. As our readers are doubtless aware, in mammals spines are nothing more than specially modified hairs, and in a porcupine the transition from a spine to an ordinary hair can be easily seen. There are many rodents in which a certain number of scattered spines are mingled with the fur of the back, but our remarks will be confined in the main to the forms in which the spines predominate sufficiently to render them the most striking feature in the external appearance of their possessors.

Commencing with the rodents, our first representatives of the spiny mammals will be the true porcupines (*Hystrix*), which are such well-known creatures as to require but brief description. These animals conform, of course, to the ordinary rodent type in having a single pair of large chisel-like incisors in each jaw ; and their spines are most developed on the middle line of the head and back, the hinder part of the body, and on the short tail. Whereas, however, those on the body are solid throughout and pointed at each end, the spines at the extremity of the short tail are in the form of hollow quills inserted by narrow stalks. It is these hollow quills that make the loud rattling sound heard when a porcupine is walking ; and it appears to us not improbable that they may have given rise to the old legend of the porcupine ejecting its spines when attacked, as such hollow quills might well have been thought to be receptacles for the ordinary spines. Although their owner is unable to voluntarily eject the latter, their pointed bases render them easily detached, and leopards which habitually feed on porcupines are found to be actually bristling with their quills. In attacking its foes, the porcupine rushes at them backwards, and thus gives full effect to its weapons. All the members of the typical genus are characterized by their large size, short tails, and highly convex skulls, and are confined to the warmer regions of the Old World. The brush-tailed porcupines (*Atherura*) from West and Central Africa and the Malayan region are, however, of much smaller size, and also distinguished by their much longer tails, which terminate in a brush of flattened spines, and are thus evidently less specialized creatures.

America is tenanted by a group of porcupines easily distinguished from their Old World cousins by having the soles of their feet covered with rough tubercles, instead of being perfectly smooth, and also by their comparatively short spines being mingled with a number of long hairs, by which they may be partially concealed. The Canada porcupine (*Erethizon*) differs from all the other American species in having a short stumpy tail, and also in its non-arboreal habits ; its spines being almost hidden by the hairs. In parts of North America these porcupines are so abundant as to be a positive nuisance, and an enterprising engineer, with true American "cuteness," hit upon the original idea of utilizing their bodies as fuel for his engine —apparently with the most satisfactory results. The lighter-built tree-porcupines (*Synetheres*), which are mainly characteristic of the southern half of the American continent, are easily distinguished by their long tails, which, as in so many South American mammals, are prehensile. These porcupines are thoroughly arboreal in their habits, and it is therefore easy to understand why their spines are so much shorter than those of their terrestrial Old World cousins, who have to rely solely on these weapons for their protection.

In addition to the members of the porcupine family, there are several other groups of rodents which develop a more or less complete coating of spines. Among the most remarkable of these groups are the spiny mice (*Acomys*) of Syria and Eastern Africa, one of which, when it has its spines erected, is almost indistinguishable at the first glance from a diminutive hedghog. The spiny rat of Celebes (*Echinothrix*) is another member of the mouse family having the fur thickly intermingled with spines. In a third rodent family (*Octodontidæ*), nearly all the members of which are South American, there is also a genus (*Echinomys*), taking its name from the number of flattened spines mingled with the fur of the back characterizing all its representatives. It will thus be obvious that even in a single mammalian order we have several

instances where a protective coat of spines must have
been acquired quite independently.

This independent origin is still more clearly indicated
when we come to the consideration of the hedgehog and its

Fig. 16.—The Hedgehog.

allies, which bear precisely the same systematic relationship to the porcupines as is presented by the true moles to
the mole-voles, as described in our last chapter. That is to
say, whereas the hedgehogs and true moles belong to the
insectivorous order, the porcupines and the mole-voles
are herbivorous rodents. In spite, then, of the general
similarity of appearance between a hedgehog and a porcupine, or, still better, a spiny mouse, we shall find, as
already mentioned, that whereas the two latter have the
ordinary chisel-like rodent teeth, the former has several
narrow and somewhat irregularly-shaped teeth in the
front of the jaws, while its back teeth are crowned with
numerous sharp cusps, instead of having nearly smooth
grinding surfaces. Accordingly, from the purely systematic
point of view there is no justification for calling the

common hedgehog the "British porcupine"; but, on the other hand, if we allow similarities in external appearance to be our guide in nomenclature, there are just as good grounds for applying the latter title to the hedgehog as there are for giving the names of golden or Cape mole, sand-mole, mole-vole, and marsupial mole to four of the creatures noticed in the last chapter. Our ancestors, to whom the hedgehog was commonly known as the "urchin," went, however, a step further than this, and, from the resemblance of its spine to those of the mammal, gave the name of sea-urchin to the *Echinus*, a title which has stuck to it ever since. Systematic zoologists need not then wax so wroth as we have known them do when the name of "British porcupine" is applied to the urchin, seeing that the analogies of nomenclature are sufficient to justify its use. As the sea-urchins come most distinctly under the title of spiny animals, it may be mentioned here that, although the spines of the common British species are not unlike those of the hedgehog, yet their structure is totally different. Thus, whereas the spines of the mammal are of a horny nature, those of the invertebrate owe their solidity to the presence of carbonate of lime, and always break with the characteristic oblique fracture of the mineral calcite. Moreover, whereas the spines of the hedgehog are implanted in its skin, those of the sea-urchin are entirely external, being movably attached by their hollow bases to knobs on the surface of the shell or "test." Whether our worthy ancestors believed that the land and sea-urchins were connected by ties similar to those which in their estimation affiliated barnacle-geese to barnacles, or how the name "urchin" came also to indicate a child, we are quite unaware.

In place of terminating in sharp points, by which they are but loosely attached to the skin, like those of the porcupine, the spines of the hedgehog terminate interiorly in small knobs, which are placed beneath the skin, and may thus be compared to pins stuck through a piece of soft leather. Beneath the skin lies a layer of muscle known as the panniculus carnosus; and it is by the action of this muscle on their heads that the spines are raised from a recumbent to a vertical position when the creature rolls itself up into a ball—an action of which all porcupines are quite incapable. Not only does the hedgehog differ from the porcupine in this respect, but it is likewise peculiar in using its spines as a means of protection when throwing itself down a vertical bank or precipice, and by this means is able to accomplish a vertical descent of over a dozen feet without the slightest harm. As regards the development of its spiny armour, the hedgehog is perhaps the most highly specialized of all the spiny mammals, in spite of the inferior length of its spines as compared with those of the porcupine. Hedgehogs are now represented by about a score of species ranging over Europe, Africa, and a considerable portion of Asia; while the existing genus dates from the middle portion of the Miocene division of the Tertiary period. That the spines characterizing the existing forms have been independently developed within the limits of the group is pretty conclusively indicated by the close affinity of the hedgehogs to the long-tailed and spineless Malayan insectivores known as gymnuras; fossil types apparently indicating an almost complete transition from the gymnuras to the hedgehogs. In case anyone should suggest that the latter, and not the former, might be the ancestral stock, it may be mentioned that while the gymnuras have a generalized type of dentition and long tails, the hedgehogs have the teeth much reduced in number and specialized in character, and short tails.

In Madagascar the place of the hedgehogs is taken by an entirely different group of insectivores known as the tenrecs, all of which have a certain number of spines mingled with the fur, at least in the young condition, and some of which are so hedgehog-like in general appearance that by the non-zoological observer they would certainly be regarded as members of the *Erinaceidæ*. The tenrecs differ, however, from the hedgehogs precisely in the same manner as the golden moles differ from the true moles. That is to say, whereas in the latter the crowns of the upper molar teeth are quadrangular, with their cusps arranged in a somewhat

W-like manner, in the former these teeth are triangular, with their cusps arranged in a V. There are five species of tenrecs, classed under three generic headings, and all characterized by the absence or small size of the tail. The largest, and at the same time the most generalized of all, is the common tenrec (*Centetes ecaudatus*), which attains a length of from twelve to sixteen inches, and is characterized by the absence of a tail, by the rows of spines on the back being shed in the adult state, and also by certain peculiar features in the dentition which appear to indicate relationship with the pouched mammals. The spiny tenrecs (*Hemicentetes*) are much smaller animals, of the size of moles, in which the longitudinal rows of spines on the back are retained throughout life. They have the same number of teeth as fully adult individuals of the common tenrec ; but whereas in the latter there are four upper molars and two upper incisors, in the spiny tenrecs there are three of each of these teeth. In the loss of the last molar these tenrecs are evidently more specialized than the common species, but the presence of the third incisor shows that they are descended from a still more generalized type. Lastly, we have the hedgehog-tenrecs (*Ericulus*), in which the whole upper surface of the body, as well as the short tail, is thickly beset with spines, thus giving the hedgehog-like appearance from which the creatures derive their name. The dentition is more reduced than in either of the upper groups, thus indicating the greater specialization of the genus ; although the presence of a short tail indicates direct descent from a tailed ancestor. Although it is quite clear that the three genera of tenrecs are divergent branches from a common stock, yet it is not impossible that they may indicate the manner in which the complete coat of spines characterizing the third group has been gradually evolved. Against this view it may, however, be urged that if the common tenrec indicated the first commencement of the spiny coat, it would be more likely to find the spines in the adult rather than in the young. Be this as it may, the wide difference between the hedgehogs and the tenrecs, coupled with the affinity of the former to the gymnuras, leaves no doubt that the spines have been acquired independently in the two groups.

As we found the last representative of the mole-like animals among the pouched mammals, so we observe that spiny animals are represented among the still lower egg-laying mammals by the spiny anteaters, or echidnas, of

FIG. 17.—The Spiny Anteater.

Australia and New Guinea. The two kinds of echidnas differ, however, externally from all the spiny animals hitherto mentioned in the production of the muzzle into a long, toothless tubular beak; and their spines are short, and in some cases largely concealed by the fur. Neither of them have the power of rolling the body into a ball; and, whereas the Australian echidna has five toes to each

foot, in the Papuan species the number is generally reduced
to three. The wide structural differences separating the
egg-laying mammals from all other members of their class
render it almost unnecessary to observe that the spines
of the echidnas are an entirely independent development.
Like so many of the spined mammals, the echidnas have
extremely short tails and thick bodies, with the neck
indistinctly marked. We have thus decisive evidence
that a more or less complete coat of spiny armour has
been independently acquired in the following groups of
mammals, viz., in the porcupines, mice, and octodonts
among the rodents; in the hedgehogs and tenrecs among
the insectivores; and in the echidnas among the egg-
layers. From the perishable nature of these appendages
we have, unfortunately, no evidence as to the existence of
spines among fossil mammals; but from the foregoing
considerations, we are strongly inclined to think they may
be mainly characteristic of later epochs.

In this connection it is interesting to notice that the
spiny globe-fishes (*Diodon*, &c.), often termed "sea-
hedgehogs," in which the spines are bony and therefore
capable of preservation, do not date back below the
Tertiary, and that spiny fishes are unknown in earlier
epochs. Moreover, some—although by no means all—of
the palæozoic sea-urchins appear to have had very minute
spines. Hence it would rather seem as though the
history of spines has been exactly the opposite of that of
bony armour, which has tended gradually to disappear with
the advance of time.

Finally, we have to notice the general similarity in
appearance of so many of the more specialized spiny
mammals, due not only to their bristly coat, but likewise
to the general shortness or absence of the tail, and the
rounded, plump form of the whole body. The bearing of
this independent development of spines in so many groups
of mammals, together with the acquirement of a general
external resemblance in the creatures thus clothed, on
questions of wider import, forms the subject of the next
chapter, in which we shall also have to take into consideration the conclusions reached in the previous one.

CHAPTER V.

PARALLELISM IN DEVELOPMENT.

In the course of the three preceding chapters, it has been to a great extent our aim to show that certain animals may resemble one another very closely in general external appearance, or may possess certain peculiar structural features in common, without being in any way intimately related; thus rendering it evident that such similarities of form or structure have been independently acquired, and are not inherited from a common ancestor. It has been shown, for instance, that mammals belonging to several distinct orders, or to different families of the same order, may assume such a marked external resemblance to the common mole as to be designated in popular language by the same general title; while in other cases a more or less striking approximation to the type of the ordinary hedge-hog has resulted from the independent development of very similar spines in totally distinct groups. Furthermore, it has been pointed out in the first of the three chapters that large tusks of very similar form may be independently developed in the jaws of totally different distinct groups of mammals; and even that certain extinct reptiles have acquired tusks which are almost indistinguishable from those of some of the carnivorous mammals, there being no direct relationship between the members of these groups.

With regard to the external similarities of form in the above-mentioned instances, we have seen reason to believe that their inducing cause has been either similarity of habit or the need of protection; while in the case of the tusks the similarity may in certain instances be due to the necessity for efficient offensive weapons. Be their causes, however, what they may, it is evident that such resemblances among animals, which are, so to speak, accidental, indicate what may be termed a kind of parallel development ; and as such parallelism in development, or shortly "parallelism," has recently attracted a good deal of attention among biologists, we propose in this chapter to present to our readers some of the more striking instances of this feature. In the instances above alluded to, the parallelism is either to a great extent shown in external characters, or in structures which are easily modified ; but, as we shall indicate in the sequel, in other cases it affects deeply-seated structures ; and its inducing cause is then very difficult to surmise on any of the ordinarily accepted doctrines of evolution. It will, moreover, be obvious that the acceptation of parallelism in development—and accepted to a certain extent it must undoubtedly be—throws a new difficulty in the interpretation of the affinities of animals, since before saying that an identity in some structural feature between the members of any two groups indicates their relationship, we have first of all to determine whether such similarity of structure is due to parallelism, or is inherited from a common ancestral type. As is so generally the case when any new theory is started, a host of enthusiastic writers have welcomed parallelism with acclamation, and have attempted to apply it in a number of instances where there is not at present sufficient evidence of its existence. We have been told, for instance, that the American monkeys have no relationship with their reputed cousins of the Old World ; that whalebone whales are not allied to the sperm-whale and dolphins ; that cats have no kinship with civets or other modern carnivores ; and that the egg-laying mammals have been evolved from a reptilian or amphibian stock, quite independently of all other members of the mammalian class—their resemblances being solely due to parallelism. At present, we confess, we are totally unable to accept any of these conclusions, and we think it somewhat improbable that if the members of either of the above-mentioned pairs of groups had an independent origin, they would have presented such a similarity, both externally and internally, as we find to be the case. Still, however, it must be borne in mind that there is a considerable amount of evidence that the modern

horses of the Old World and their extinct cousins of the
New have been independently derived from earlier horse-
like animals; and if this be confirmed, it will be one of
the most remarkable known instances of parallelism, and
will tend to show that many others may exist. Turning
from these more or less problematical cases, we proceed to
notice *seriatim* certain well-marked instances of parallel
development as to the existence of which there can be no
doubt ; commencing with those displayed in external
features, and then referring to such as are more deeply
seated.

As regards external resemblances, it is, of course—and
more especially so far as the lower animals are concerned
—not always easy to distinguish between parallelism and
mimicry. The above-mentioned instances of mole-like and
hedgehog-like animals belong, however, clearly to the
former category. The external resemblances existing
between swifts and swallows, which have no sort of
relationship to one another, likewise comes under the same
head. Another striking instance is to be found in the
assumption of a snake-like form, accompanied by abortion
of the limbs, by certain lizards, such as the familiar
English "blind worm," it being very doubtful, to our
thinking, if this can be explained by mimicry. Still more
remarkable is the external resemblance presented by
whales and dolphins to fishes, and by the extinct fish-
lizards (Ichthyosaurs) to both ; such resemblances being
clearly due to parallel development induced by the needs
of adaptation to a purely aquatic life. If, moreover, it
should prove that the whalebone whales have no connection
with the dolphins, we should then have a far more remarkable instance of this feature, and one extending to internal
structures as well as to external form. A similar remark
will also apply to the case of the true seals and the eared
seals, which it has been suggested may be of independent
origin. The independent acquisition of wings by birds,
bats, and the extinct flying dragons, or pterodactyles, may
likewise certainly come under the heading of parallel
development, so far as external form and the adaptation
to a particular mode of life are concerned ; although here.

as in the instance of whales and fishes, the structural
features, by the aid of which the adaptation has been
brought about, are very different in the respective groups.

In respect to the teeth and dentition, many very well-
marked instances of parallelism may be adduced. We
have already referred to the similarity of the tusks in different groups of mammals; while in a previous chapter
we noticed the loss of upper front teeth, not only in the
modern ruminants, but likewise in the peculiar even-toed
ungulate known as *Protoceras*, of which the skull is figured
on page 53. The rhinoceroses likewise show a gradual
tendency to the loss of front teeth, resulting in the African
forms in the disappearance of the whole series from both
jaws. More remarkable, however, is the tendency to a
complete loss of the whole of the teeth in certain groups
of all the higher classes of vertebrates, as exemplified by
all modern birds, by turtles and tortoises, as well as certain
of the extinct flying dragons and fish-lizards among
reptiles, and by the great anteater, the scaly anteater,
and the echidnas, or spiny anteaters, among mammals;
in all of which teeth are completely lacking. Moreover,
in many of these animals, such as birds, tortoises, and
probably the toothless flying dragons, the beak-like jaws
thus produced were sheathed in horn, thus showing a
kind of double parallelism, viz., the loss of teeth coupled
with the acquisition of a horny sheath to the jaws.

Another instance of parallel development afforded by
teeth relates to the gradual heightening of the crown and
the production of a flat plane of wear not only among
several distinct groups of hoofed mammals, such as the
elephants, horses, and ruminants, but likewise among the
rodents. It is evident that these high-crowned teeth
have been independently evolved in the three great groups
of hoofed mammals, of which the above-named creatures
are typical representatives; and it may be added that the
molars of horses and ruminants present a further evidence
of parallelism in their assumption of a more or less
decidedly crescent-like (selenodont) structure, the essential
peculiarities of which are described in the eighth chapter.
This resemblance between the molars of horses and

E

ruminants is, however, comparatively remote, but in the latter group and the camels these teeth are so alike as to require an expert to distinguish between them. Nevertheless, there is good evidence to show that the camels and the ruminants, if not also the chevrotains, have acquired their crescent-like molar teeth quite independently of one another, and it therefore yet remains for those writers who explain evolution by some mode of what they are pleased to call natural selection to account adequately for the similarity thus existing between structures of such totally different origin, when they could have been made equally efficient if unlike.

Passing from the consideration of teeth to that of limbs, we may mention the remarkable similarity displayed in the mode by which the lower segments of the limbs of the even-toed and odd-toed hoofed mammals have been gradually elongated by the formation of a cannon-bone and the disappearance of either three or four of the lateral digits; the cannon-bone in the horse consisting of but a single element carrying one digit, while in the ruminants it comprises two united elements supporting a pair of toes. This is clearly a case of parallelism in development attained by a slightly different modification of principle. The parallelism does not, however, stop here, since an essentially similar type of cannon-bone has been produced in birds; only that in that group (with the exception of the ostrich) three long bones enter into its composition, which is further complicated by the addition of a bone from the ankle above. Seeing that in all these groups the parallelism has been arrived at by a different structural modification, the explanation of its mode of evolution is much less difficult than in the case of the molars of the camels and ruminants, where, as we have seen, the structure is practically identical.

Recent discoveries in North America have brought to light the existence of a kind of secondary parallelism among certain peculiar mammals which may be included among the hoofed or ungulate division of that class. In the even-toed group of that division, as exemplified by the pigs and ruminants, it is the third and fourth digits

which are symmetrical to one another, and tend to
develop at the expense of the others ; while in the odd-
toed group it is the third or middle digit which is
symmetrical in itself and undergoes an ultra development.
Now there are certain extinct creatures, which, while
having claws instead of hoofs at the ends of their toes,
yet are so closely related to the hoofed mammals that

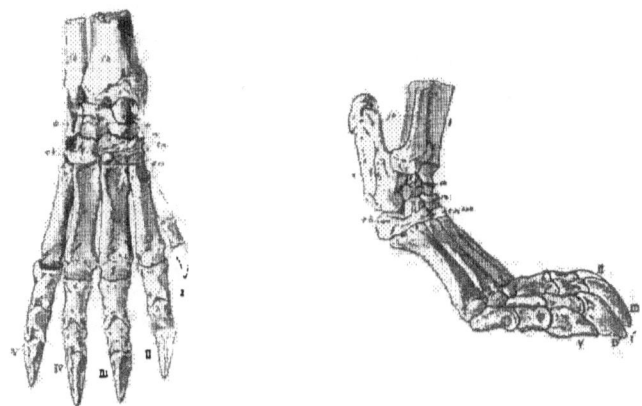

Fig. 18.—Front and side views of the Hind Foot of *Artionyx*.
(After Osborn.)

their separation therefrom is almost impossible. In one
of these, which has long been known in Europe as the
chalicothere, the third digit, as in the odd-toed hoofed
mammals, is symmetrical in itself, and larger than
either of the others ; whereas in the newly-described
animal known as *Artionyx* the third and fourth digits
resemble those of the even-toed division of the hoofed
mammals in being larger than the others and symmetrical
to a line drawn between them. While, therefore, we have
clearly a parallel development on different lines between
the odd and even-toed hoofed animals in the development
of a cannon-bone, these extinct clawed, hoof-like mammals
(for want of a better expression) show a parallel parallelism (to coin another expression) which, if it had continued,
might have resulted in the development of a two-clawed

ruminant and a two-clawed horse. To our thinking, this is indeed one of the most curious phases of development yet discovered.

In addition to the parallelism in the evolution of their molar teeth and limbs, the hoofed mammals likewise exhibit the same feature in regard to the second vertebra of the neck. As many of our readers are probably aware, the first, or atlas vertebra of the neck, turns with the head when the latter is moved sideways, the axis of rotation being formed by a process—the odontoid process arising from the second or axis vertebra, and projecting into the central hollow of the atlas. Now in pigs and likewise all the primitive hoofed mammals, the so-called odontoid process is (as in ourselves) in the form of a flattened peg. On the other hand, in the ruminants, the modern horses, and the camels, which, as we have seen, represent three distinct *phyla* of the order, this peg has become modified into a spout-like half-cylinder, which must clearly have been separately evolved in each of these three groups. It is true that such a half-cylinder affords a far better basis of support for a heavy skull than does a mere peg; but the curious part of the matter is why these half-cylinders should be so exactly alike in the different groups, seeing that, as it would not be difficult to design some other structural modification by which the same end might have been attained, there is no necessity for their similarity.

Our last example of parallelism will be drawn from two groups of extinct North American hoofed mammals, to which brief allusion has already been made in the chapter on "Tusks and their Uses"; the one group being known as uintatheres, while the second is represented by a single species to which the name of *Protoceras* has been applied. Now, although the uintatheres have five-toed feet approximating in structure to those of elephants, while in *Protoceras* each foot approximated to the ruminant type, in both groups the skull, as shown in the accompanying figure, was armed with several pairs of large, irregular, bony processes, which during life may have been sheathed in horn; while in each case a pair of long tusks projected from the upper jaws, which were totally devoid of front or incisor teeth. Had such skulls been discovered without any indication as to the nature of the limbs with which they were associated, they would inevitably have been assigned to the same group of animals. The resemblance existing between them, is, however, clearly due to parallel development, and we are thus shown another striking instance of caution necessary in endeavouring to determine the affinities of extinct animals from the evidence of incomplete remains.

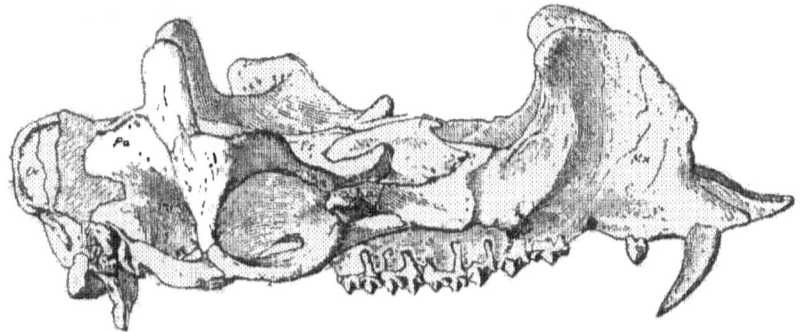

Fig. 19.—Skull of *Protoceras*. (After Osborn.)

Our last instance of parallelism is alluded to in a later chapter, entitled "The Oldest Fishes and their Fins," in the course of which it is shown that while both the most ancient birds and the oldest fishes had long tapering tails with the joints of the backbone gradually diminishing in size, and each carrying either a pair of feathers or a pair of fin-rays, in all the modern representatives of the former group, and in a large section of the latter, the end of the vertebral column has been aborted into a composite bone, from which either the feathers or the rays of the tail diverge in a fan-like manner.

In conclusion, we may say that although there is no very great difficulty in satisfactorily accounting for external parallelism obviously due to the necessity for adaptation to a particular mode of life, or in explaining those instances where a particular result has been brought about by

different methods, yet when we find precisely similar
structural modifications in different groups of animals
which clearly cannot be traced to a common ancestry, and
for which equally efficient substitutes could be readily
suggested, we are fain to confess that the ordinarily
accepted explanations of evolution appear to us altogether
inadequate. Putting this aspect of the matter aside, our
readers will, however, see from the imperfect sketch given
above, what an important factor in evolution parallel
development really is, and how largely it is likely in the
future to modify our present views as to the mutual
relationships of animated nature.

CHAPTER VI.

TOOTHED WHALES AND THEIR ANCESTRY.

WITHIN the entire limits of the great mammalian class, there are, perhaps, no creatures which arouse a larger amount of interest—both among the general public and among naturalists—than those included under the names of whales, dolphins and porpoises, and collectively known as cetaceans. One reason for this universal interest is, doubtless, that among these denizens of the deep are comprised the largest animals, not only of the present day, but likewise, so far as our information allows us to speak, of all epochs. Then, again, the fact that such apparently fish-like creatures are really warm-blooded mammals, suckling their young in the manner distinctive of all other members of the class, and being under the necessity of coming to the surface at certain intervals for the purpose of breathing, cannot fail to strike even the most unobservant mind as being something quite beyond the ordinary. Moreover, the momentary glimpses which in general are all we obtain of these animals, and the halo of mystery which still to a great extent enshrouds their mode of life, are likewise important elements in generating the widespread interest they arouse. To the zoologist, cetaceans are indeed not only of prime importance as being the sole mammals which have assumed a purely fish-like form, and have become so thoroughly adapted to a completely pelagic life as to be unable to exist on land, but their study gives rise to many problems as to their origin and relationships, and the mode in which they attained their present condition.

Into the consideration of the leading external features of cetaceans we need not enter very fully, merely pointing out that while the general contour of the body is fish-like, the tail-fin, or flukes, differs essentially from that of a fish in being horizontal instead of vertical ; while

in place of the two sets of paired fins characterizing a
fish, a whale has but a single pair of flippers representing
the greatly modified fore limbs of other mammals.
The hind limbs have, indeed, been completely lost
externally, although more or less imperfect traces of
them may still be detected deeply imbedded among the
muscles of the body. In the great majority of the group
the back is furnished with an upright fin, very similar in
appearance to the unpaired back-fin of a fish. Whereas,
however, such a back-fin is constantly present in fishes,
in cetaceans it may be absent or present in different species
of the same genus; while if we were to cut through such a
fin we should find a total absence of the slender spine-like
bones characterizing those appendages in a fish; a similar
condition also obtaining in the flukes. In marked contrast
to the scaly armour of the majority of modern fishes, the
skin of a cetacean is for the most part completely naked;
although the frequent presence, in the young state at
least, of a few scattered bristles in the region of the mouth
is of itself sufficient to indicate the derivation of these
strangely modified creatures from more ordinary mammals.
As regards their coloration, we may again, however, note
a similarity to most pelagic fishes, in that while the upper
parts are generally dark, the lower surface of the body is
of a light hue; this arrangement being, of course, designed
to render all these animals as inconspicuous as possible
when viewed in the water either from above or from below.
Although the flippers show no external indications of
toes, and are unprovided with nails, yet their internal
skeleton comprises the same elements as occur in the
limbs of any ordinary mammal, and is thus quite different
from that of a fish. This structural similarity is, however,
to a certain degree obscured by the alteration in the form of
the bones, and also by the circumstance that the number
of joints in the skeleton of the individual toes is increased
beyond the normal. As external ears would be mere
useless incumbrances, these appendages are absent; while
the aperture of the ear itself is reduced to an extremely
minute size. To prevent the ingress of water during the
periods of submergence, the apertures of the nostrils, which

may be either double or single, can be completely closed at will, and are only opened at such times as the creatures come to the surface to breathe, when a column of water is generally thrown up by the rush of expired air let loose

Fig. 20.—The Bridled Dolphin. (From True, *Bull. U. S. Nat. Museum.*)

shortly before the head reaches the surface. In all their internal structures, as well as in the mode of production and nourishing of their young, cetaceans conform strictly to the ordinary mammalian type ; and we accordingly see that their assumption of a fish-like form is, with the exception of the loss of the hind limbs and the modification of the front pair into flippers, mainly superficial. In departing from the fish-type in having the expansion of the tail-fin horizontal instead of vertical, the necessity of having an organ capable of bringing them rapidly to the surface has been the inducing cause ; while in order to prevent their blood from being reduced below the proper temperature by the chill of the surrounding water, the whole body is invested with a thick layer of oily fat, commonly known as the blubber, beneath the skin.

Existing cetaceans are divisible into two great groups, distinguished as the whalebone whales and the toothed whales ; the latter group including sperm-whales, together with the grampuses, porpoises, and dolphins. The most obvious distinction between these two groups, as the terms applied to them indicate, relates to the presence or absence in the adult condition of true teeth.

Confining our attention in the present chapter to the toothed cetaceans, of which the dolphins and their allies are the least specialized representatives, we find that the two jaws may be in some cases provided with a full series of teeth, while in other forms the number of teeth may be reduced to a single pair, or even, as in the male narwhal (Fig. 9), to a solitary tusk. Whether, however, the teeth be many or few (and in the female narwhal there are none of any functional importance), the structure known as whalebone is never developed in the mouth; while all the members of the group are further distinguished from the whalebone whales by the circumstance that the nostrils invariably open by a single external aperture, which is very frequently in the form of a transverse crescentic slit, closed by an overhanging valve. In the latter respect these cetaceans are more specialized than are the whalebone whales; and as the presence of teeth in the former indicates that they could not have been

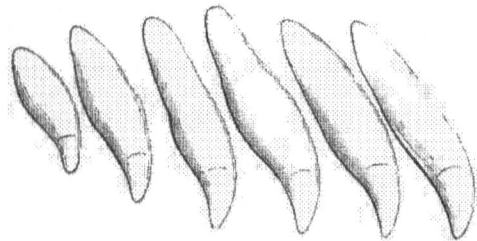

Fig. 21.—The last six Upper Teeth of the Killer. (After Sir W. H. Flower.)

derived from the latter, it is evident that the two groups are of extreme antiquity, and have undergone a parallel development. Till recently, it has indeed been considered that they were divergent branches from some common ancestral type; but, as mentioned in the chapter on " Parallelism in Development," it has been lately suggested that each may have had a totally distinct origin, although the evidence in favour of such a view is, at present at least, far from conclusive.

In the structure of their teeth, the modern toothed whales differ very widely from the generality of mammals. In the first place, their teeth are always of the simple structure shown in Fig. 21, having conical or compressed crowns, and undivided roots; while, secondly, there is only one single series developed, the replacement of the anterior ones characterizing the majority of mammals being wanting. From this simple structure of their teeth it has been argued that these cetaceans are among the most primitive of all mammals; but, altogether apart from the conclusive evidence that all whales (as proved by their breathing air) are derived from land mammals, it has recently been shown by the researches of Dr. Kükenthal, of Jena, that this view is quite untenable. By examining embryos of young cetaceans of this group, that observer has demonstrated that there are actually rudiments of a second series of teeth, which, although never coming to maturity, serve to show that there were once two complete sets, and that the permanent teeth correspond, in part at least, to the milk-teeth of other mammals; thus indicating that the present state of the cetacean dentition is a degraded one. Hitherto it has not, indeed, been shown by embryology that the teeth of this group were originally of a complex type (although in the case of the whalebone whales this has been demonstrated), but, fortunately, here palæontology comes to our aid. Thus in the middle of the Tertiary period there occur remains of what may be termed shark-toothed dolphins (squalodonts), in which the permanent teeth are differentiated into distinct series, corresponding to the incisors, canines, premolars, and molars of other mammals; while, for all we know to the contrary, there may also have been a regular replacement of the more anteriorly placed teeth. In these shark-toothed dolphins the molar teeth, instead of being of the simple structure of those represented in our illustration, were severally implanted in the jaws by two perfectly distinct roots; while their large, laterally compressed, and somewhat fan-

shaped crowns were furnished with a number of cusps on
their hinder cutting-edges. Indeed, these teeth much
resemble the premolar teeth of a dog or the molars of a
seal; and they obviously serve to indicate a transition from
the modern toothed cetaceans towards ordinary mammals.
This, however, is by no means all, since in a still earlier
portion of the same division of the Tertiary period there
occur other cetacean-like animals known as zeuglodonts,
which have still more complicated teeth, and otherwise
depart further from the modern cetacean type—so much so,
indeed, that they have been regarded by some writers as
more nearly allied to the seals. In our own opinion they are,
however, undoubtedly primitive cetaceans, and thus serve,
not only to connect the present group with other mammals,
but also, in conjunction with the shark-toothed dolphins, to
show that the simple teeth of the former are clearly produced by degeneration from a complex type. As regards
the particular group of land mammals from which whales
were derived, it has been thought that their nearest allies
are with the ancestors of the pig-like hoofed mammals.
This, indeed, is the view of Sir W. H. Flower; but we
confess that from the nature of the teeth of the two extinct
groups above mentioned, coupled with certain resemblances
of the skeleton of the zeuglodonts to those of seals, we
are rather more inclined to look among flesh-eating land
mammals for the lost ancestors. Still, however, it must
be remembered that in the early Eocene period, which is
probably the very latest epoch at which cetaceans could
have originated, the distinction between carnivorous and
hoofed mammals was but imperfect, so that, after all, the
ancestral cetacean stock may well have been of an
extremely generalized type. At present, however, we are
almost completely in the dark in all that concerns this
interesting subject.

By far the largest of all the toothed whales is the
gigantic sperm-whale, the typical representative of a family
characterized by the absence of teeth in the upper jaw of
the adult, while those of the lower jaw are very variable
both as regards form and number. In the sperm-whale, of
which the male attains a length of between fifty and sixty

feet, the lower teeth vary in number from fifty to sixty on each side, and are characterized by their large size and pointed crowns, upon which there is not a trace of enamel. Another characteristic feature of this animal is the enormous size of the head, which terminates in an abruptly truncated muzzle of great depth, and in a cavity of which is contained a peculiar oily substance, yielding when refined the well-known spermaceti. An even more valuable product of this animal is ambergris, which, while accumulated as a concretion in the intestine, is generally found floating on the surface of the sea ; in appearance it is an amber-coloured substance, containing a number of the horny beaks of the squids on which the sperm-whale subsists.

Omitting mention of the lesser sperm-whale—the only other member of the first division of the family—we pass on to mention the bottle-nosed and beaked whales, characterized by having all the lower teeth, with the exception of a single pair, rudimentary, and concealed in the gum. In the bottle-noses—so named from the extreme convexity of the crown of the head of the adult males, which rises suddenly above the short beak—there is but a single pair of teeth, situated in the front of the lower jaw, and even these are invisible during life. These whales, of which there is but a single well-defined species (*Hyperoödon rostratus*), although not exceeding some thirty feet in length, are valuable on account of their oil, as well as from yielding spermaceti from their skulls. In contradistinction to the sperm-whale, they carry, in common with the beaked whales, a large back-fin. In the beaked whales, of which there are three existing genera, the skull is produced into a long beak, of which the upper half is formed by a solid bone of ivory-like density ; while in the lower jaw there are either one or two pairs of teeth, which, although variable in position, are generally of large size. It is one of these whales (*Mesoplodon layardi*) which is alluded to in the chapter on " Tusks and their Uses," as being provided with teeth of such a size as actually to impede the free opening of the mouth. From their general avoidance of the neighbourhood of coasts, and their apparently somewhat solitary habits, extremely little is known of the life-history of the beaked whales, of which the skeletons are but poorly represented in our museums. At the present day very rarely seen in the English seas, during the Pliocene period these whales must have been extremely numerous in the North Sea, since their fossilized beaks are amongst the most common vertebrate fossils obtained from the crags of Suffolk and Essex. The same deposits, together with others of corresponding age on the Belgian coasts, have also yielded remains of a number of extinct whales, more or less closely allied to the sperm-whale, thus indicating that the latter is the last survivor of a once numerous group. The teeth of many of these fossil sperm-whales differed, however, from those of their living cousins in having their crown capped with enamel.

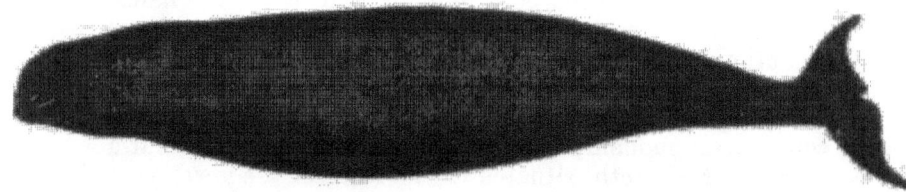

Fig. 22.—The Indian Porpoise. (From True, *Bull. U.S. Nat. Museum.*)

In the general presence of numerous teeth in both the upper and the lower jaws, the dolphins and their allies the porpoises, killers, and white-whale differ from all the above-mentioned forms, and thus constitute a second family —the *Delphinidæ*. The group is a numerous one, which is split up into a number of genera, some of which are by no means easy to distinguish. They may, however, be roughly ranged under two main divisions, in one of which the muzzle is short and rounded, as in the porpoises and black-fish, while in the other, as represented by the dolphins (Fig. 20), it is produced into a longer or shorter beak, of which the base is marked off from the main portion of the head by a distinct re-entering angle. The most aberrant member of the first group is the spotted Arctic narwhal, of which a brief notice, accompanied by a figure, was given in the chapter on "Tusks." Allied to the narwhal is the beautiful white-whale (*Beluga*), which is likewise a northern species, distinguished by its glistening white skin, the absence of any tusk, and the presence of numerous well-developed teeth in the fore part of the jaws; neither of these species having a back-fin.

Although the name "porpoise" is applied indiscriminately to several members of the family, it should properly be restricted to a few comparatively small-sized species characterized by the presence of some twenty-five small and flattened teeth with spade-shaped crowns on either side of each jaw. Porpoises are among the most common and familiar of all cetaceans, their rolling gambols being well known to all who have made a voyage; but whereas the common porpoise has a distinct back-fin, in the species represented in our illustration that appendage is lacking. Less familiar, on the other hand, are the much larger and handsomely coloured killers, or grampuses (*Orca*), differing by the great size of their teeth (Fig. 21), which are usually twelve in number on each side, and the great development of the back-fin.

FIG. 23.—Pacific Black-Fish. (From True, *Bull. U. S. Nat. Mus.*)

Attaining a length of about twenty-five feet, the killer derives its title from its rapacious habits; a single specimen having been known to swallow several whole seals in succession, while not unfrequently several individuals have been observed to combine their forces to attack and kill the larger members of the order. Killers may always be easily recognized while swimming near the surface by the great height of their nearly vertical back-fins. Allied cetaceans with smaller teeth (*Orcella*) frequent the Bay of Bengal and ascend some distance up the Irrawadi. Another well-marked type is the black-fish (Fig. 23), characterized by its remarkably short and rounded head, the uniformly black hue of the skin, and by the eight or twelve small and conical teeth being confined to the front portion of the jaws.

There are other less well-known representatives of this group which we have not space to notice ; and we accordingly pass on to say a few words about dolphins—a term which should be restricted to those forms having a distinctly marked beak. Since, however, sailors will persist in speaking of dolphins indifferently, either as bottle-noses or porpoises, the inexperienced landsman must be on his guard not to confound them when thus spoken of either with the bottle-nosed whales or the true porpoises. Dolphins, which are divided into numerous genera, according to the number of their teeth, the length of the beak, and other characters, are all comparatively small species, seldom exceeding some ten feet in length ; and while the great majority are marine, a few ascend some of the larger tropical rivers, such as the Amazon. Fish of various kinds constitute their usual prey ; but one peculiar species recently described from the Cameruns district is believed to subsist on sea-weed. Of the better-known types, the common dolphin represents the genus *Delphinus*, the bottle-nosed dolphins constitute a distinct genus (*Tursiops*), while the long-beaked dolphins are separated as *Steno*.

To the foregoing group of beaked dolphins the ordinary observer would, doubtless, be disposed to refer three peculiar species severally restricted to the larger rivers of India, the Amazon, and the mouth of the La Plata river, but as these differ more or less markedly from other dolphins in certain structural features they are referred to a distinct family. Moreover, since these peculiarities approximate to a more generalized type, while their fresh-water habits and scattered distribution indicate extreme antiquity, it is not improbable that these three dolphins.

are the most primitive of all existing cetaceans. At present we have, indeed, no evidence of fossil forms allied to the susu or Gangetic dolphin (*Platanista*); but in the older Tertiary deposits, both of the United States and Europe, there occur the remains of dolphins evidently nearly allied to the two existing South American species (*Inia* and *Stenodelphis*), and thus clearly proving the antiquity of those types. It may be added that it is probable that the ancestors of the cetaceans which first took to an aquatic life were inhabitants of fresh-water, and it is therefore only what we should expect that the most primitive of the existing representatives of the order were likewise of fluviatile habits. On the other hand, it must be confessed that the similarity in structure of the widely separated existing fresh-water dolphins is somewhat difficult to account for on this hypothesis.

CHAPTER VII.

WHALEBONE AND WHALEBONE WHALES.

SEEING that the substance so-called has nothing in common with true bone, many zoological writers object to the use of the term "whalebone"; and they have accordingly proposed the substitution of the word "baleen," which has been specially coined for the purpose.   To our thinking, there is, however, no necessity for this substitution of a word of foreign origin for such a well-known English name, any more than there is for replacing the native term "black lead" by its foreign equivalent "graphite."   Everybody knows what is meant by whalebone or black lead, while comparatively few are familiar with the terms "baleen" and "graphite"; and as the two former are every bit as good as their foreign substitutes, we prefer to employ them. If, indeed, there should exist any persons so misguided as to imagine either that whalebone is equivalent to the bone of whales, or that black lead has any sort of affinity with lead, we fear that the substitution of the terms "baleen" and "graphite" would not much aid in removing their ignorance.

The substance which we accordingly take leave to call whalebone is one of the chief essential characteristics by which the whalebone whales are distinguished from the toothed whales forming the subject of the preceding chapter.   As our readers are probably aware, this substance is attached to the upper surface of the mouth of the whale, from which it depends in the form of a series of parallel, narrow, elongated, triangular plates, placed transversely to the long axis of the mouth, with their external edges firm and straight, but the inner ones frayed out into a kind of fringe.   The longest plates of whalebone are situated near the middle of the jaw, from which point the length of the plates gradually diminishes towards the two extremities, where they become very short. There is, however, whalebone and whalebone, and whereas in the Greenland whale the length of the longest plates varies from some ten to twelve feet, while the total number of plates in the series is about 380, in the great rorquals or fin-whales the length is only a few inches, while the number of plates is considerably less. To accommodate the enormous whalebone plates of the Greenland whale, the bones of the upper jaw are greatly arched upwards, while the slender lower jaw is bowed outwards, thus leaving a large space both in the vertical and horizontal directions, the transverse diameter of the space being much wider below than above. When the mouth is

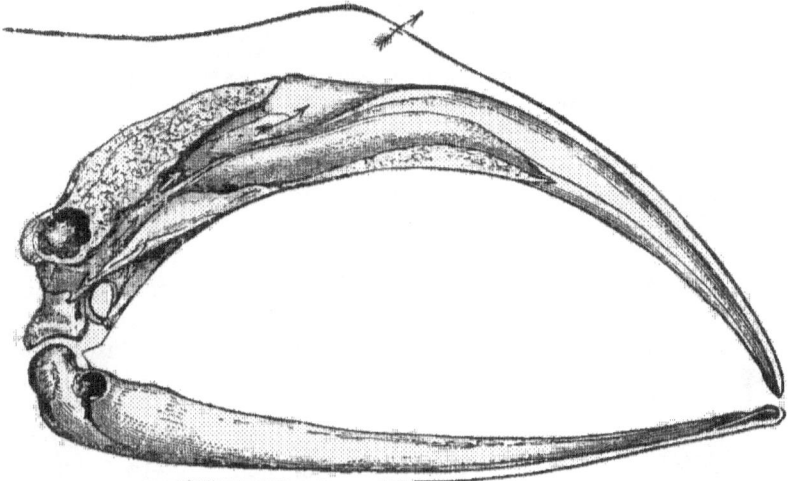

FIG. 24.—Section through the Skull of the Greenland Whale, with the outline of the soft parts. The arrows indicate the position of the nasal passage and aperture.

closed, the plates of whalebone are folded obliquely backwards, with the front ones lying beneath those behind them ; but directly the jaws are opened, the elastic nature of this wonderful substance causes it to spring at once into a vertical position, and thus form a sieve-like wall on both sides of the mouth, the thin ends of the plates being prevented from pushing outwards by the stiff lower lip which

overlaps them. By elevating its enormous fleshy tongue
within the cavity thus formed, the whale causes the
enclosed water to rush out between the plates, leaving such
small creatures as it contained lying high and dry on the
surface of the tongue ready for swallowing.

In structure, whalebone (which, by the way, although
black in the Greenland whale, is white in some of the
other species) is of a horny nature, and grows from
transverse ridges on the mucous membrane of the roof of
the mouth ; being, in fact, nothing more than an ultra-
development of the ridges on the palate of a cow, hardened
and lengthened by an excessive growth of a horny superficial or epithelial layer. The whole of this stupendous
horny growth takes place, however, after birth, young
whales having smooth palates, with no trace of the horny
plates. Although at birth young whalebone whales show
no traces of the substance from which the group derives
its name, they equally exhibit no evidence of the
presence of teeth. If, however, their jaws be examined at
a still earlier stage of development, it will be found that
there are a number of small teeth lying within a groove
beneath the gum on each side of both the upper and the
lower jaws. Previous to birth these teeth become absorbed,
and thus never cut the gum. Their presence in this
transitory stage is, however, of the deepest interest to
the evolutionist, since they unmistakably indicate the
derivation of the whalebone whales from ancestors
provided with a full series of functional teeth. This,
however, is not the whole of the story these rudimentary structures have to tell. From the recent
investigations of Dr. Kükenthal, it appears that in addition
to the above-mentioned tooth-germs, the jaws of very
young whales likewise exhibit traces of a still earlier
deciduous series of milk-teeth ; thus showing that the
former correspond to the permanent series of other
mammals. Accordingly, these tooth-germs do not represent
the functional teeth of the toothed whales, which, as we
have seen in the chapter on that group, correspond to the
milk-teeth of ordinary mammals. Even more remarkable
are certain observations relating to the structure of these

tooth-germs. It has been shown, indeed, that the teeth in the latter part of the series, when first formed, consist of a number of adjacent cusps, and that as development proceeds these cusps become completely separated from each other so as to constitute distinct individual teeth of a simple conical form.

This discovery is of the very highest import, since it serves to indicate how the numerous simple conical teeth of the dolphins and other toothed whales have probably been derived by the splitting and subdivision of originally complex cusped teeth more or less closely resembling those of the extinct zeuglodonts referred to in the last chapter.

Being primarily distinguished from the toothed whales by the total absence of teeth after birth and the presence of whalebone in the adult condition, the whalebone whales present certain other distinctive characters to which we may now briefly allude.

In the first place, the whales of this group differ externally from all those furnished with teeth in that their nostrils open externally by two distinct longitudinal slit-like apertures ; while, if we cut into the head, we shall find that there is a distinct organ of smell, of which all traces have disappeared among the toothed whales. Moreover, instead of the skull being invariably unsymmetrical in the region of the nose, as it is in the latter group, it retains the normal symmetry ; while, instead of the mere nodules which in the toothed whales represent the nasals of other mammals, in the whalebone whales these bones are fairly well developed. Then again, the lower jaw of any member of the present group may always be distinguished from that of a toothed whale not only by the absence of teeth, but likewise by the circumstance that each of its branches is much bowed outwards in the middle, while their anterior extremities are connected together merely by ligamentous tissue, instead of by a bony symphysis of greater or lesser length. Many other points of difference between the two groups might be cited, but we have especially referred to those mentioned above, for the reason that while the presence of whalebone indicates that in one respect the

members of this group are more specialized than their
toothed cousins, in regard to the structure of the skull in
the region of the nose and the retention of an organ of
smell, as well as in the double apertures of the nostrils,
they depart less widely from the ordinary mammalian
type.   And it is from this evidence that zoologists regard
the two main groups of whales as being widely divergent
branches from a common ancestral stock, if, indeed, they
do not go so far as to consider that each has had a totally
independent origin.

With regard to the number of forms by which they are
represented, the whalebone whales are far less numerous
than the toothed group, the whole of them being included
within a single family.   What the group lacks in number
is, however, amply made up by the great bodily size of
its various representatives.   Among the toothed whales,
the   sperm-whale   alone   attains gigantic dimensions;
whereas in the present group there are several species,
such as the Greenland whale, which nearly equal that
enormous creature in bulk, while two of the rorquals far
surpass it.   This, however, is by no means all, for whereas
the great majority of the dolphins and porpoises are
relatively small cetaceans of less than twenty feet in
length, the smallest member of the present group is the
southern pigmy whale, which attains a length of twenty
feet, while next to this comes the lesser rorqual, whose
length is frequently close on thirty feet.

Of the various genera of this group, the most specialized
are the typical or right whales (*Balæna*), in which the
black whalebone is far longer and more elastic than in any
of the others, except the pigmy whale ; while, in order to
accommodate it in the mouth, the skull has the
palate narrower and much more highly arched, and
the two branches of the lower jaw more outwardly
bowed than in the other members of the group.   Externally, these whales—which are commercially of far
greater value than the undermentioned rorquals—are
characterized by the inordinately large dimensions of the
head, by the smooth throat, the moderate length of the
flippers, and the total want of a back-fin.   So gigantic

indeed, is the size of the mouth in these whales that its capacity actually exceeds that of the whole of the other cavities of the body ; and yet the size of the throat is so small as almost to justify the common nautical saying that a herring will choke a whale. There are but two well-defined species of right whales, viz., the Greenland or Arctic whale (*B. mysticetus*), and the southern right whale (*B. australis*), the latter of which was nearly exterminated some centuries ago in the Atlantic by the Basque whalers, while the former is only too likely to share the same fate at the hands of their modern successors in the Arctic seas. Of the two, the Greenland whale is decidedly the more specialized, having a much larger head and longer whalebone, and is in this respect *facile princeps* among its tribe ; although, as it only measures from forty-five to fifty feet in length, it is inferior in point of size to the rorquals. The skeletons of both these whales are characterized by the whole of the vertebræ of the neck being welded together into a solid mass ; and the same feature is exhibited by those of the pigmy whale (*Neobalæna*) of the southern seas, which, as already mentioned, is a mere dwarf among giants, as it does not exceed some twenty feet in total length. Agreeing also with the right whales in its smooth throat, the pigmy whale differs by having a small hook-like fin on the back, while its long and elastic whalebone is white instead of black. Far larger than the last, the great grey whale of the Pacific (*Rhachianectes*) forms a kind of connecting link between the right whales and the rorquals, having the smooth throat and finless back of the former, while its whalebone is even shorter and coarser than in the latter ; the palate consequently is but little vaulted, and the entire head smaller in proportion to the body.

The remaining whales of this group are divided into hump-backs (*Megaptera*) and rorquals or finners (*Balænoptera*) ; both of which are characterized by the presence of a number of parallel groovings or flutings in the skin of the throat, as well as of a back-fin (whence their name of finners or fin-whales), and also by the shortness and coarseness of their whalebone, which is generally of a yellowish colour. Their flukes are, moreover, less expanded than are those of the right whales ; while, as already said, their heads are relatively smaller and lower, with the cavity of the mouth much less vaulted. In their skeletons, the vertebræ of the neck differ from those of the right whales in being longer and completely disconnected from one another ; and in this respect the Pacific grey whale holds a position intermediate between the two groups. Hump-backs, of which there is but a single species, are specially characterized by the shortness and depth of the body, which behind the shoulder rises above the level of the back-fin, and by the exceeding elongation and slenderness of the flippers, which are about equal to one-fourth the total length of the animal. The female hump-back, which somewhat exceeds her partner in size, attains a length of from forty-five to fifty feet, or about the same as that of the Greenland whale. An enormous whale, believed to belong to this species, became some years ago entangled in the telegraph cable line off the coast of Baluchistan with three turns of the cable round its body. On the other hand, in the rorquals, the body is long and slender, while the flippers are small and pointed. Of the four well-established species of this genus, the blue or Sibbald's whale (*B. sibbaldi*)—the "sulphur-bottom" of the American whalers—enjoys the proud distinction of being the largest of all known animals, whether living or extinct, attaining the enormous length of from eighty to eighty-five feet ; while the common rorqual (*B. musculus*) comes in a good second with a length of from sixty-five to seventy feet. Both the others are, however, considerably smaller.

As regards their distribution in time, whalebone whales have left their remains commonly enough in the Pliocene strata of all parts of the world, and they likewise occur in those of the Miocene period ; but, although a single vertebra from the Eocene beds of Hampshire has been assigned to a member of this group, there is at present no decisive evidence that they had come into existence at such an early date. Since most of the Pliocene species are of smaller dimensions than their living representatives, it appears that these marvellous creatures had only attained their maximum size shortly before the time when their very existence was to be threatened by the relentless hand of man.

Formerly the only members of this group of whales which were thought worthy of general pursuit were the right whales; the shortness of their whalebone, coupled with their relatively small yield of oil, and their tremendous speed, rendering the rorquals scarcely worth the trouble and risk of hunting. Of the two right whales, the Greenland species, as being the more abundant, received the greatest share of attention; and so relentless has been its pursuit, that it is now either well-nigh exterminated from many of its ancient haunts, or has retreated still farther north to regions almost impossible of access. As showing how the constant persecution in the Greenland seas has told upon the size of the comparatively few remaining individuals of this species, it may be mentioned that the eleven specimens killed there during the season 1890-91 yielded an average of less than 8 cwt. of whalebone, whereas in five taken during the same season in Davis Strait the average yield was more than double this amount. In consequence of this diminution in the number and size of the Greenland whale, the value of whalebone has of late years gone up enormously; and whereas some time ago whalebone of over six feet in length sold at £1000 per ton, in 1892 it had reached the enormous price of upwards of £2800 per ton. The southern right whale yields a smaller quantity of rather shorter and less valuable bone, now selling at from £1600 to £1800 per ton; the quantity obtained from a well-grown example varying from 800 to 1200 pounds. The amount of oil produced by a whale of the same species averages from eight to fourteen tons, of which the present market value is about £28 per ton. If the unfortunate animals are not allowed some respite, it is only too probable that the supply will before long cease altogether.

As another result of this growing scarcity of the Greenland whale, attention has been directed to the previously despised rorquals and hump-backs; the employment of steam whaling vessels, and the use of explosive harpoons, having enabled the whalers easily to cope with the greater speed of these cetaceans.   At Hammerfest, a special " fishery " has, indeed, been established for the capture of rorquals, where the take of these animals is now large. The products of these whales are, however, nothing like the value of those of the Greenland species; and if the latter, together with the southern right whale, be so nearly exterminated as to render pursuit no longer profitable, the supply of long whalebone will absolutely come to an end.

Postscript.—Since the foregoing was written, the author has discovered that certain fossil toothed whales from Patagonia had their nasal bones nearly as well developed as in the whalebone whales ; this discovery removing one of the difficulties in regarding the latter as the descendants of the former,

CHAPTER VIII.

RUMINANTS AND THEIR DISTRIBUTION.

FROM early times we find the function of ruminating, or "chewing the cud," recognized as a peculiarity of the group of mammals known in semi-popular language as ruminants. In "Deuteronomy," for instance, the animals permitted for food are those that "chew the cud and part the hoof"; while the swine, "which part the hoof but do not chew the cud," are forbidden. On the other hand, the camel, which chews the cud but has not paired hoofs, is in the forbidden list. In the permitted animals we thus have a recognition of the group of ruminants as represented by oxen, sheep, and deer; of which no better short definition can be given than that they chew the cud and have each foot furnished with a pair of hoofs symmetrical to a vertical line between them. The want of the paired hoofs in the camels, which are also cud-chewers, show, however, that these two characteristics will not hold good for the entire group. As we proceed, we shall find that there are structural features, common to the group, in addition to the peculiarity of rumination; but before going further, we may observe that the recognition of their paired hoofs, coupled with the absence of rumination, is an exact statement of the relationship of the swine to the true ruminants.

The word "ruminant" comes from the Latin *rumen*, which was applied both to the "cud" and to that part of the stomach in which the latter is contained previous to chewing. The Greeks had a word *meruko* or *merukizo* (from *meruo*, to revolve), to express this action of cud-chewing, and a derivative from the former was used by Aristotle to designate ruminants, who thus first distinguished the group by a definite name. This early recognition of the ruminants as a group is probably due to their importance to man, the Biblical record showing

that they yielded the only mammalian food permitted to
the Hebrew, and this pre-eminence as a source of food has
scarcely decreased to the present day.   They are, moreover,

FIG. 25.—Skeleton of the Extinct Irish Deer—a typical Ruminant.

now the dominant type of larger mammals, as witness the
herds of bison which lately roamed over the American

prairies, and the droves of antelopes on the African veldt.

Commencing with the function of rumination, we may observe that it is a re-mastication of grass or other vegetable food, swallowed almost as soon as plucked, and transferred to a special receptacle in the stomach. From this it is regurgitated into the mouth by a reversed action of the muscles of the throat, and, after having undergone mastication,—or rumination—is transferred to the digesting part of the stomach. Now, it is evident that this complicated arrangement, so different from that of other animals, must be of some special advantage to the ruminants. As a matter of fact, these animals, like other large herbivores, are obliged to consume a vast quantity of food to obtain sufficient nutriment; and it is obvious that if this food had to be masticated as soon as plucked, the operation of feeding would be very protracted; but by the arrangement mentioned the requisite amount of food can be gathered within a comparatively short time, and the animals can then retire to ruminate in concealment. It is superfluous to comment on the advantage of this arrangement to creatures which, like many ruminants, have but little means of defending themselves against carnivorous foes; but we may mention that many still further increase this advantage by feeding only at dawn or evening, when they are far less conspicuous than in the mid-day glare. There is, moreover, evidence that when ruminants first appeared, this rapid feeding was of more importance than at the present day, since while many of the modern larger forms, like oxen, antelopes, and deer, are provided with formidable weapons in the shape of horns or antlers with which they can keep foes at bay, in earlier times such weapons were either absent or but feebly developed.

Seeing, then, that the function of rumination is correlated with a special compartment of the stomach for the temporary reception of the freshly-gathered food, it would be expected that animals thus provided would also possess an efficient masticating arrangement for reducing their food to the condition in which it yields the fullest nutriment. Such, indeed, is the case, the grinding-teeth

of ruminants being of a complex structure, unknown
elsewhere ; the last three of these teeth in the upper jaw
(Fig. 26) being composed of four columns, of varying

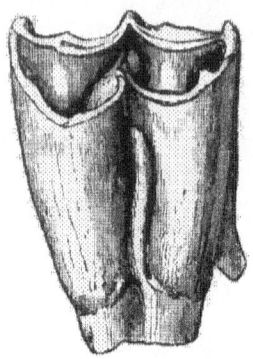

height, of which the two inner ones
are crescent-shaped ; such a type of
tooth being termed selenodont. The
lower grinding-teeth having their
crescents directed the opposite way to
those of the upper jaw, and both
upper and lower teeth consisting of
layers of different hardness, we can
scarcely imagine a better masticating
machine than is presented by the
opposition of the two series of
grinding - teeth of these animals.
Bearing in mind this structure, the
definition of cud-chewing, selenodont
mammals will suffice to distinguish
the ruminants from all other animals.

FIG. 26. — A left
Upper Cheek-Tooth of
an Antelope.

When, however, we say that these characteristics distinguish them from all other mammals it must be added
that this refers only to those of the present day, for a
transition can be traced through extinct forms to the more
simply constructed grinding-teeth of the swine. We have
already seen how the Mosaic law recognized the similarity
in the structure of the hoofs of the ruminants and the
swine, and it is curious that while under the Cuvierian
system of zoology these two groups were widely sundered,
modern palæontological researches have shown that they
are really closely related, the want of the power of chewing
the cud, with the correlated absence of the selenodont
structure of the teeth, being the chief essential features
in which the latter differ from the former.

Here a curious problem is presented to those who put
their faith in a mode of evolution dependent only upon
so-called natural causes, in that it is impossible to give
any adequate explanation of what advantage would be the
development of an incipient selenodont structure in the
teeth of the early swine-like ungulates, or at what precise
stage the function of chewing the cud, with the con-

comitant development of a separate compartment in the
stomach, was superadded to the normal mode of feeding
characteristic of the swine.

Here we must say a few words as to the structure of
the ruminant foot. The "cloven
hoof" of ruminants and swine has
become such a proverbial expression
that the idea may still linger that
this is due to the fission of a single
hoof, like that of a horse. Nothing
could, however, be further from the
truth ; the two hoofs of a ruminant
(Fig. 27) corresponding to the
terminal joints of our own middle
and ring-fingers (or the corresponding toes), which are the third and
fourth of the typical series of five.
The lateral or spurious hoofs (not
shown in Fig. 27) of the ruminants
represent our own index (second)
and little (fifth) fingers, or toes. It
is a further peculiarity of the true
ruminants and camels that the two
separate bones which in the swine
connect the two large digits with
the wrist or ankle are fused into a
single cannon-bone (Fig. 27) ; the
primary dual origin of which is
indicated by the two distinct pulley-
like surfaces at the lower end,
which carry the bones of the digits.
The peculiar little ruminants known
as the chevrotains—of which more
anon—retain, however, evidences of
their kinship with the swine, in
that some of them have the two
elements of the front cannon-bone
—or metacarpals as they are then
called—quite separate from one
another. Indeed, in the same

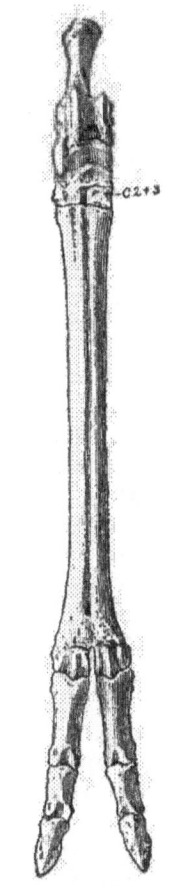

Fig. 27.—Bones of the
Hind Foot of a Rumi-
nant. The letters in-
dicate the lower bones
of the ankle. (After
Osborn.)

manner as we may trace a transition from the selenodont teeth of the ruminants to the bunodont ones of the swine, we may mark how the two-toed and cannon-boned ruminants passed into swine-like animals with four toes supported by as many separate metacarpal bones.

Having now mentioned the leading characters of a modern ruminant, as distinct from other mammals, we may refer to a peculiarity, which, although by no means characteristic of all, is a striking one, and one sharply differentiating the group from all others. This is the tendency to the development of appendages on the skull, arranged in a pair at right angles to its longer axis, and taking the form either of solid branching antlers, as in the deer, or of hollow sheaths of horn covering bony cores on the skull, as in the oxen and antelopes.

Passing to the consideration of the various kinds of cud-chewing mammals, we find that the true ruminants, or those with hoofs, no upper front teeth, and a cannon-bone in both limbs, arrange themselves in several minor groups. The most important to man are the "hollow-horned ruminants," such as oxen, sheep, goats, and antelopes, all of which are characterized by the presence of horns, at least in the males. The variety of form assumed by the horns renders this group one of the most attractive of all animals; and we have but to recall the curved and smooth horns of the oxen, the equally massive but wrinkled ones of the wild sheep, those of the ibex with their knotted points and scimitar-like backward sweep, the spear-like form of those of the gemsbok, and the spiral twist of those of the kudu and eland, to realize the variety of contour assumed by these appendages.

The oxen (including bison and buffaloes) are, with the exception of the American bison, Old World types, and were formerly abundant in Europe, where, however, they are now only represented by the bison preserved in the forests of Lithuania and the Caucasus, and by the half-wild cattle of Chillingham and some other British parks, which have been thought to be the direct descendants of the British wild ox, or aurochs, of Cæsar's time, but are more probably derived from ancient domesticated breeds which have reverted to a nearly wild state. True wild oxen now exist only in India and the adjacent regions, while wild buffalo occur both in India and Africa.

Equally characteristic of the Old World are wild sheep and goats, the "big-horn" being an outlying North American type. Both groups are essentially mountain animals, the head-quarters of the former being the highlands of Central Asia, while on the southern flanks of the same mountain-barrier the latter are more abundant. Both are also represented in the mountains of Europe; but in peninsular India there is but the wild goat of the Nilgiries, while in the whole of Africa we have only the wild sheep of Barbary and the ibex of Abyssinia. This absence of sheep and goats from Africa may, perhaps, be due to the fact that these animals are of comparatively late origin, and were probably poorly represented at the time when the other ruminants entered that continent from the north. The musk-ox of Arctic America is an aberrant form allied to the sheep.

The antelopes have a distribution nearly the reverse of that of the sheep and goats, the great majority being restricted to Africa, where there are probably fully ninety species, against about a score in all the rest of the world, except Arabia and Syria, of which the fauna is allied to that of Africa. Indeed, the only typical antelopes found beyond these regions are the black-buck, the nilgai, the four-horned antelope of India, the saiga of Tartary, the chiru of Tibet, and several members of the widely distributed gazelles. The rings marking the horns of the latter (Fig. 28) and many other antelopes are very distinctive of the group, although by no means universal. The European chamois, the goat-antelopes of India and China, and the Rocky Mountain goat of America, serve to connect the typical antelopes with the goats, and it is these alone which represent the group in Europe, to the eastward of India, and in North America. Seeing that in Tertiary times, antelopes of African types occurred in southern Europe and India, it is difficult to determine why the group should have so dwindled or disappeared there; although we can readily account for their extraordinary development when they once obtained an entry into Africa, on account of the immense area open to them, in which there was no competition by any other ruminants except buffaloes and giraffes.

To the zoologist Africa is indeed a country characterized by the number of animals living there which have disappeared from other regions; and there is no better instance of this survival than the giraffe, a ruminant that, as regards its cranial appendages, stands midway between the hollow-horned group and the deer.* We are all familiar with the ungainly and yet beautiful form of the giraffe; but it is probably less well known that giraffes once roamed over Greece, Persia, India, and China, where, as in Africa at the present day, they were accom-panied by ostriches and hippopotami. And here again we are confronted by the problem how to account for the disappearance from regions apparently exactly suited to their habits, of all these animals. The giraffe is, how-ever, not only the sole survivor of several extinct species of its own kind, but it likewise represents a lost

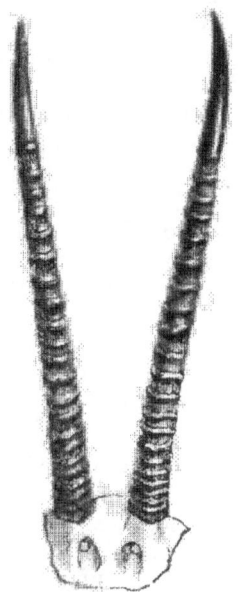

Fig. 28.—Horns of Gazelle. (From Günther.)

group of ruminants, intermediate between the horned and antlered Old World types. The head-quarters of this group was India, where, among other forms, occurs the gigantic sivathere, rivalling the elephant in bulk, and characterized by its two pairs of horns (Fig. 29), of which the hindmost were branching and antler-like, although apparently never shed, and probably covered during life with skin and hair.

If our attention has been turned to Africa as the head-quarters of antelopes and giraffes, it must be directed to

---

* See the following chapter.

other regions when we come to the deer, since, with the
exception of the Barbary stag, there is no representative of
the group in all that continent. With few exceptions, deer
are characterized by the antlers of the males, the reindeer
alone having these appendages in both sexes. They are
the only true ruminants found in South America, where
many of the species have comparatively simple antlers, and
thus show affinity with the early fossil types, some of
which were antlerless. Allied species range through North
America, but it is not till the north of that continent is
reached that we find in the wapiti a representative of our
own red deer. The red deer group extends through Europe

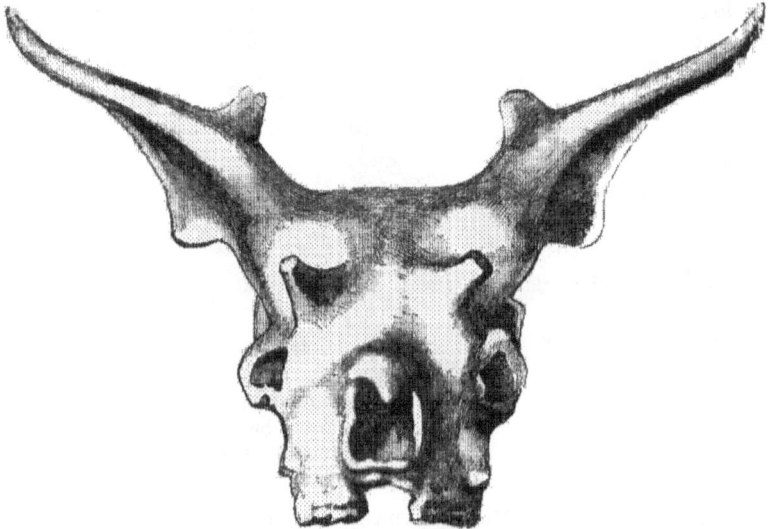

Fig. 29.—Skull of Sivathere, from the Pliocene of India.

and a large part of Central Asia, but in India and the
Malayan region it is replaced by the rusine deer, like the
sambar, in which the antlers (Fig. 30, *a*) lack the bez-tine
of the red deer (*ibid, b*). Other marked varieties of antler
are exhibited by the elk, the fallow deer, and the reindeer;
but none of these approach those of the extinct Irish deer
(Fig. 25), which may have an eleven feet span from tip to

G 2

tip. It is noteworthy that in a few small deer in which the males have no antlers, they are compensated by having long tusks in the upper jaw.

The tiny oriental chevrotains, and the larger African water-chevrotain form a group quite distinct from all the above, and are in some respects related to the swine. None of them have antlers, and the African species is the only living ruminant in which the two elements of the front cannon-bone remain separate, thus affording another instance of the survival of primitive forms in Africa.

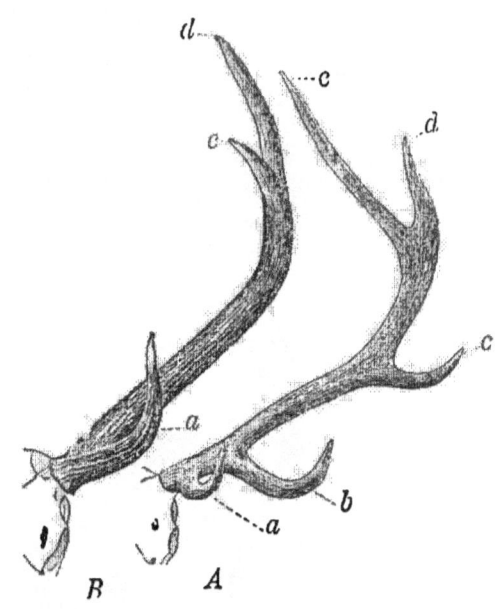

FIG. 30.—Antlers of Red (A) and Sambar (B) Deer. a brow-, b bez-, c trez-tine, d e sur-royals. (After Blanford.)

Lastly, we have the group of camels and llamas, which differ from other ruminants in that their feet form cushion-like pads, while their upper jaws possess front teeth. According to the latest researches it is considered probable that this group has diverged from primitive swine-like animals quite independently of the true ruminants, an inference which is very remarkable, showing that selenodont teeth, a complex stomach, the function of rumination, and the single cannon-bone, have been acquired quite independently in the two groups. The present distribution of camels and llamas is remarkable, the former being confined to Africa and Asia, and the latter to South America. Here, however, geology comes to our aid, for in former times camel-like ruminants were abundant in North America, while the fossil camels of India show certain resemblances to the llamas, and we can thus understand how the present distribution of the two sections of the group has come about. With the possible exception of some herds of the Bactrian species in Central Asia, which are, however, probably descended from individuals escaped from captivity, wild camels are now unknown, and we cannot even determine the original habitat of the single-humped species.

Thus ends our brief survey of the chief groups of living ruminants and their distribution. Did space permit, we might go on to refer to their extreme importance to man, both as sources of food and of clothing, and as beasts of draught and burden, but having reached our limits, we trust that we may have aroused in our readers an interest in these highly specialized animals which may induce some of them to devote further consideration to the subject.

CHAPTER IX.

THE TALLEST MAMMAL.

COMPARED with their extinct allies of earlier periods of the earth's history, it may be laid down as a general rule that the large animals of the present day are decidedly inferior in point of size. During the later portion of the Tertiary period, for instance, before the incoming of the glacial epoch, when mammals appear to have attained their maximum development, there lived elephants alongside of which ordinary individuals of the existing species would have looked almost dwarfs, while the cave-bear and the cave-hyæna attained considerably larger dimensions than their living representatives, and some of the sabre-toothed tigers must have been decidedly larger than the biggest African lion or Bengal tiger. Again, the remains of red deer, bison, and wild oxen, disinterred from the cavern and other superficial deposits of this country, indicate animals far superior in size to their degenerate descendants of the present day; while some of the extinct pigs from the Siwalik Hills of northern India might be compared in stature to a tapir rather than to an ordinary wild boar. The same story is told by reptiles, the giant tortoise of the Siwalik Hills, in spite of its dimensions having been considerably exaggerated, greatly exceeding in size the largest living giant tortoises of either the Mascarene or the Galapagos Islands. The Siwalik rocks have also yielded the remains of a long-snouted crocodile, allied to the garial of the Ganges, which probably measured from fifty to sixty feet in length, whereas it is very doubtful if any existing member of the order exceeds half the former of these dimensions. If, moreover, we took into account totally extinct types, such as the megatheres and mylodons of South America, and contrasted them with their nearest living allies—in this instance the sloths and anteaters—the discrepancy in size would be still more marked, but such a comparison would scarcely be analogous to the above.

Fig. 31.—The Giraffe.

To every rule there is, however, an exception, and there are a few groups of living large mammals whose existing members appear never to have been surpassed in size by their fossil relatives. Foremost among these are the whales, which, as we have seen in a previous chapter, now appear to include the largest members of the order which have ever existed. The so-called white, or square-mouthed, rhinoceros of South Africa seems also to be fully equal in size to any of its extinct ancestors ; and the same is certainly true of the giraffe, which may even exceed all its predecessors in this respect. Whether, however, the fossil giraffes, of which more anon, were or were not the equals in height of the largest individuals of the living species, there is no question but that the latter is by far the tallest of all living mammals, and that it was only rivalled in this respect among extinct forms by its aforesaid ancestors. Moreover, if we exclude creatures like some of the gigantic dinosaurian reptiles of the Secondary epoch, which, so to speak, gained an unfair advantage as regards height by sitting up on their hind legs in a kangaroo-like manner, and limit our comparison to such as walk on all four feet in the good old-fashioned way, we shall find that giraffes are not only the tallest mammals, but likewise the tallest of all animals that have ever existed.

In the great majority of animals that have managed to exceed all their kin in height, the increment in stature has been arrived at by lengthening the hind limbs alone, and thus making them the sole or chief support of the body. In some of these cases, as among the living kangaroos and the extinct dinosaurs, the body was raised into a more or less nearly vertical position, and the required height attained without any marked elongation of the neck. In birds, on the other hand, like the ostrich, the body is carried in nearly the same horizontal position as in a quadruped, but both the hind legs and the neck have been elongated. The giraffe, however, has attained its towering stature without any such important departure from the general structure characterizing its nearest allies, and thus preserves all the essential features of an ordinary quadruped. Belonging, as we have had occasion to mention

in the preceding chapter, to the great group of ruminant
ungulates, among which it is the sole living representative
of a separate family, the giraffe owes its height mainly to
the enormous elongation of two of the bones of the legs,
coupled with a corresponding lengthening of the vertebræ
of the neck. As in all its kindred, the lower segment of
each leg of this animal forms a cannon-bone, the nature
of which has been explained in the same chapter ; and
in the fore limb it is the bone below the wrist (commonly termed the knee), and the radius above the latter,
which have undergone an elongation so extraordinary as
to make them quite unlike, as regards proportion, the
corresponding elements in the skeleton of a ruminant,
such as an ox, although retaining precisely the same
structure. Similarly, in the hinder limb, it is the cannon-
bone below the ankle joint or hock, and the tibia or shin-
bone above, which have been thus elongated. To anyone
unacquainted with their anatomy, it might well appear
that a giraffe and a hippopotamus would differ greatly
in regard to the number of vertebræ in their necks ; but,
nevertheless, both conform in this respect to the ordinary
mammalian type, possessing only seven of such segments.
Whereas, however, those of the latter animal are very
broad and short, in the giraffe they are extremely long and
slender, attaining in full-grown individuals a length of
some ten inches. This remarkable adherence to one
numerical type in the neck vertebræ is, indeed, a very
curious feature among mammals ; the extreme contrasts
in respect of form being exhibited by those of the Greenland whale, in which each vertebra is shortened to a broad
disc-like shape, and the giraffe, where it is equally narrow
and elongated.

As regards the height attained by the male of the tallest
of quadrupeds, there is, unfortunately, a lack of accurate
information, and since it is probable that the majority of
those now living are inferior in size to the largest
individuals which existed when the species was far more
numerous than at present, it is to be feared that this
deficiency in our knowledge is not very likely to be
remedied. By some writers the height of the male giraffe

is given at sixteen feet, and that of the female at fourteen feet, but this is certainly below the reality. For instance, Mr. H. A. Bryden states that a female he shot in southern Africa measured nearly seventeen feet to the summits of the horns, while a male measured within less than half an inch of nineteen feet; Sir S. Baker, whose experiences are derived from the north-eastern portion of the continent, also asserting that a male will reach as much as nineteen feet. From the evidence of a very large, though badly preserved specimen in the Natural History Museum, it may also be inferred that fine males certainly reach the imposing height of over eighteen feet.

Although this towering stature is the most obvious external feature of the giraffe, it is not one which would of itself justify the naturalist in classing the animal as the representative of a family apart from other ruminants; and we must accordingly inquire on what grounds such separation is made. On the whole, the most distinctive structural peculiarity of the giraffe is to be found in the nature of its horns. These, as mentioned in the preceding chapter, are quite unlike those of any other living ruminant, and take the form of a pair of upright bony projections arising from the summit of the head in both sexes, and completely covered during life with skin. In the immature condition separate from the skull, these horns become in the adult firmly attached to the latter; while below them, in the middle of the forehead, is another lower and broader protuberance, sometimes spoken of as a third horn. Obviously, these horns—for want of a better name—are quite unlike the true horns of the oxen and antelopes, or the antlers of the deer; and this essential difference in their structure is alone quite sufficient to justify the reference of the giraffe to a family all by itself. When, however, we come to inquire whether the creature is more nearly akin to the deer or to the hollow-horned ruminants (as the oxen, antelopes, and their allies are termed), we have a task of considerable difficulty. Relying mainly on the structure of its skull, and its low-crowned grinding teeth, which are invested with a peculiar rugose enamel having much the appearance of the skin of the common black slug, some naturalists speak of the giraffe as a greatly modified deer. A certain justification for this view is, indeed, to be found in the circumstance that the liver of the giraffe, like that of the deer, is usually devoid of a gall-bladder. Occasionally, however, that appendage, which is so characteristic of the hollow-horned ruminants, makes its appearance in the giraffe, thus showing that no great importance can be attached to it one way or another. On the other hand, in certain parts of its soft anatomy, the creature under consideration comes very much closer to the antelopes and their kin than to the deer. It would appear, therefore, on the whole, that the giraffe occupies a position midway between the deer on the one hand and the antelopes on the other; while as neither of these three groups can be regarded as the direct descendant of either of the other two, it is clear that we must regard all three as divergent branches from some ancient common stock.

In general appearance, the giraffe is too well known to require description, but attention may be directed to a few of its more striking external peculiarities. One remarkable feature is the total lack of the small lateral or spurious hoofs, which are present in the great majority of ruminants, and attain relatively large dimensions in the reindeer and musk-deer. Indeed, the only other members of the whole group in which these hoofs are absent are certain antelopes; but this absence cannot be taken as an indication of any affinity between the latter and the giraffe, since it is most probably the result of independent development. Equally noticeable are the large size and prominence of the liquid eyes, and the great length of the extensile tongue; the former being obviously designed to give the creature the greatest possible range of vision, while the extensibility of the latter enhances the capability of reaching the foliage of tall trees afforded by the lengthened limbs and neck. In comparison with the slenderness of the neck, the head of the giraffe appears of relatively large size; but this bulk, which is probably necessary to the proper working of the long tongue, is compensated by the extreme lightness and porous structure of the bones of the skull. Lastly, we may note that the long tail, terminating in a large tuft of black hairs, is a feature unlike any of the deer, although recalling certain of the antelopes.

Somewhat stiff and ungainly in its motions—the small number of vertebræ not admitting of the graceful arching of the neck characterizing the swan and ostrich—the giraffe is in all parts of its organization admirably adapted to a life on open plains dotted over with tall trees, upon which it can browse without fear of competition by any other living creature. Its wide range of vision affords it timely warning of the approach of foes; from the effect of sand-storms it is protected by the power of automatically closing its nostrils; while its capacity of existing for months at a time without drinking renders it suited to inhabit waterless districts like the northern part of the great Kalahari desert. And here we may mention in passing that the camel has gained a reputation for being adapted for a desert-life above all its allies which is not altogether deserved. It is true, indeed, that a camel can and does make long desert journeys, but these can only be maintained during such time as the supply of water in its specially constructed stomach holds out, and when this fails there is not an animal that sooner knocks up altogether than the so-called "ship of the desert." Did their bodily conformation and general habits admit of their being so employed, there can indeed be little doubt that the giraffe and some of the larger African antelopes, which are likewise independent of water, would form far more useful and satisfactory beasts of burden for desert travelling than the, to our mind, somewhat over-rated camel. Returning from this digression, it must be mentioned that when we speak of the giraffe being independent of water, we by no means intend to imply that it never drinks. On the contrary, during the summer this ruminant, when opportunity offers, will drink long and frequently; but it is certain that for more than half the year, in many parts of southern Africa at least, it never takes water at all. In certain districts, as in the northern

Kalahari, this abstinence is, from the nature of the country, involuntary ; but according to Mr. Bryden, the giraffes living in the neighbourhood of the Botletli river—their only source of water—never drink therefrom throughout the spring and winter months. When a giraffe does drink, unless it wades into the stream, it is compelled to straddle its fore-legs far apart in order to bring down its lips to the required level, and the same ungainly attitude is perforce assumed on the rare occasions when it grazes.

There is yet one other point to be mentioned in connection with the adaptation of the giraffe to its surroundings before passing on, and this relates to its coloration. When seen within the enclosures of a menagerie— where, by the way, their pallid hue gives but a faint idea of the deep chestnut tinge of the dark blotches on the coat of a wild male—the dappled hide of a giraffe appears conspicuous in the extreme. We are told, however, that among the tall *kameel-dhorn* trees, or giraffe-mimosas, on which they almost exclusively feed, giraffes are the most inconspicuous of all animals ; their mottled coats harmonizing so exactly with the weather-beaten stems and with the splashes of light and shade thrown on the ground by the sun shining through the leaves, that at a comparatively short distance even the Bushman or Kaffir is frequently at a total loss to distinguish trees from giraffes, or giraffes from trees.

At the present day, it is hardly necessary to mention, the single species of giraffe is exclusively confined to Africa, not even ranging into Syria, where so many other species of animals otherwise characteristic of that continent are found. This restricted distribution was, however, by no means always characteristic of the genus ; for during the Pliocene period extinct species of these beautiful animals roamed over certain parts of southern Europe and Asia. The first of these extinct giraffes was discovered by Falconer and Cautley many years ago in that marvellous mausoleum of fossil animals, the Siwalik Hills of north-eastern India ; remains of the same species being subsequently brought to light in the equivalent deposits of Perim Island, in the Gulf of Cambay, and likewise in the Punjab. A second species has also left its remains in the newer Tertiary rocks of Pikermi, near Athens ; while those of a third have been disinterred in China, and a fourth in Persia. It was, indeed, believed for a long time that France also was once the home of a member of the genus, but the specimen on which the determination was based is now known to be a jaw-bone belonging to the existing species. Although we are, unfortunately, unacquainted with the geology of the greater part of Africa, the foregoing evidence points strongly to the conclusion that giraffes (together with ostriches, hippopotami, and certain peculiar antelopes) are comparatively recent emigrants into that continent from the north-east ; but, as we have previously had occasion to mention, the reason why all these animals have totally died out in their ancient homes is still one of the darkest of enigmas.

Unknown in the countries to the north of the Sahara, as well as in the great forest regions of the west, which are unsuitable to its habits, the giraffe at the present day ranges from the north Kalahari and northern Bechuana-land in the south, through such portions of eastern and central Africa as are suited to its mode of life, to the southern Sudan in the north. Unhappily, however, this noble animal is almost daily diminishing in numbers throughout a large area of southern and eastern Africa, and its distributional area as steadily shrinking. Whether it was ever found to the south of the Orange river and in the Cape Colony may be a moot point, although, according to Mr. Bryden, there are traditions that it once occurred there. Apart from this, it is definitely known that about the year 1818 these animals were met with only a little to the north of the last-named river ; while as late as 1886 they were still common throughout the Transvaal, and more especially near the junction of the Marico with the Limpopo river. Now their last refuges in these districts are the extreme eastern border of the Transvaal (where only a few remain), and the district lying to the north of Bechuanaland and known as Khama's country, or Bawangwato, together with the northern Kalahari. Even here, however, their existence is threatened, as there is a proposal to put down tube-wells in the waterless Kalahari, which, if successfully accomplished, will open up the one great remaining stronghold of the animal to the merciless hunter. Unless, therefore, efficient and prompt measures are taken for its protection, there is but too much reason to fear that the giraffe will ere long be practically exterminated from this part of Africa ; although, fortunately, it has a prospect of surviving for many years to come in the Sudan and Kordofan. The great majority of the giraffes killed at the present day in southern Africa are shot solely for the sake of their skins, which are now, owing to the practical extermination of rhinoceroses south of the Zambesi, and the ever-increasing scarcity of the hippopotamus, used in the manufacture of the formidable South African whips known as *jamboks*. The value of a skin usually varies, according to size and quality, from £2 10s. to £4, although they have been known to fetch £5 apiece ; and it is for the sake of such paltry sums that one of the noblest and most strange of mammals stands in imminent danger of extermination !

We may conclude this notice by mentioning that although the giraffe was familiar to the Romans of the time of the empire, by whom it was known as the camelopard, it appears to have been almost completely lost sight of in Europe in later times till the closing decades of the eighteenth century, although a single example is stated to have been exhibited alive in Florence some four centuries ago. With that exception, it seems to have been generally regarded as a fabulous animal until one was shot near the Orange river in 1777 by an Englishman, and another by the French naturalist, Le Vaillant, in 1784. From that time onwards our knowledge of the animal and its habits gradually increased, although it was not till the spring of 1836 that four living specimens from the Sudan were brought alive to London, where some of their descendants lived continuously till 1892, since which date the species has been unrepresented in the Regent's Park.

CHAPTER X.

LEMURS.

In a previous chapter we have had occasion to refer to Africa as an archaic kind of land containing types of mammalian life which, while formerly widely spread over the Old World, are now restricted to that continent. If this preservation of ancient types be highly characteristic of continental Africa, still more markedly is it so of the large island of Madagascar, lying off its eastern coast. In Africa itself many of the ancient types are more or less closely allied to other living mammals, and most of them belong to orders which are abundantly represented in other parts of the world. Very different is, however, the case with Madagascar, of which the great peculiarity is that it has preserved to us a whole fauna of those remarkable animals known as lemurs, which are represented elsewhere only in Africa and the Oriental region, and there by a comparatively small number of species belonging to genera totally distinct from those found in Madagascar. To put the matter more clearly, it may be stated that out of a total of thirteen genera of living lemurs no less than eight are absolutely confined to the island of Madagascar, while the remaining five are distributed over Africa, India, and the Malayan islands; two out of the five being African, one Indian, and two Malayan. The disproportion is, however, not even adequately expressed by the above statement, since, while most of the Malagasy genera are represented by a considerable number of species, of those found in other regions only one has more than two species, while two out of the other four have but a single species each. Lemurs are, indeed, in every sense the characteristic mammals of Madagascar, being far more common in its woods and coppices than are squirrels in those of this country. So common are they said to be in certain parts of the island that, according to the French traveller,

M. Grandidier, it is impossible to beat through a single copse without turning out at least one of these strange creatures.

In order to arrive at the true reason of the present distribution of any group of animals, it is always necessary to consult the records of geology; and it appears from these that while lemurs were unknown both in Europe and North America during the Pliocene and Miocene divisions of the Tertiary period, when we reach the upper part of the Eocene epoch we find their remains occurring in company with those of the extinct anoplotheres and palæotheres, or other allied animals, in both the eastern and western continents. It is, however, hardly necessary to observe that the whole of these early lemurs belonged to genera which are quite distinct from any of those living at the present day, although one of them appears to have nearly been allied to the African group.

The discovery of these extinct lemurs, which is but comparatively modern, at once reveals the fact that this group of animals is a very ancient one, and one formerly widely spread over the globe, and represented by at least one species in our own island. Not many years ago it was sought to explain the present peculiar distribution of lemurs by the supposition that a large island or continent formerly existed in the Indian Ocean; and the name Lemuria was suggested, appropriately enough, for this hypothetical land. From this presumed ancestral home it was considered that the lemurs had spread on all sides, some to find a refuge in the Malayan islands, others in the forests of Ceylon and southern India, but the larger number in Madagascar and Africa. Unfortunately, however, for a very pretty theory, the geologists had a word to say on the matter; and this word was to the effect that Lemuria could not possibly have existed at the time its presence was required for the needs of the theory.

With our fuller knowledge of fossil lemurs any such hypothetical land is, however, quite unnecessary to explain the present distribution of the group. At or about the time these animals existed in Europe there is little doubt that they were also spread over Africa, which there is

H

good evidence to show was formerly connected by land with Madagascar. We do not, indeed, yet know how the ancient lemurs of Europe got into Africa, but when once there it is certain that they were ultimately cut off from Europe by a sea which stretched from the Atlantic to the Bay of Bengal in upper Eocene times. For some time afterwards, during which Africa and Madagascar still had a free communication, the large mammals characteristic of Miocene Europe were unknown in the lands to the south of this great sea, where lemurs and other lowly animals flourished in security. During some portion of this period Madagascar must have become separated from Africa, while an upheaval of land once more brought Africa into connection with Europe and Asia, and thus allowed it to be overrun by the great hoofed and carnivorous mammals which had hitherto existed only to the north of the dividing sea. This incursion of large quadrupeds at once put a final stop to the supremacy of the lemurs in Africa, where only a few species have since managed to survive by the aid of their nocturnal habits. In Madagascar, however, where there are still no large quadrupeds, and where the only carnivores are certain civet-like animals, the lemurs have continued to flourish in full exuberance, and the existing state of that island thus offers to our view a picture of what must have been the condition of Africa previous to the advent of its present fauna from the north. The few lemurs now inhabiting the Indian and Malayan region are, doubtless, also survivors from the original central home of the group, which have found safety in the dense forests of the regions they inhabit.

A good deal more might be said on the subject of the past and present distribution of lemurs, but by this time the reader will probably be impatient to know something of the characteristics of the animals thus designated by naturalists. Of course comparatively little can be said on this subject in an essay of the present length, and it unfortunately happens that the lemurs, as a whole, do not present any very strongly marked single external feature by which they can be distinguished at a glance from all other mammals. It is, however, only with some of the lower monkeys of the New World that they could possibly be confounded by observant persons (although we have heard of some members of the family being mistaken for sloths) ; but the distinct geographical distribution of the two groups renders it improbable that any confusion is likely to arise between them. The nearest relations of the lemurs are undoubtedly the monkeys, and most naturalists in this country are now agreed in regarding the former as

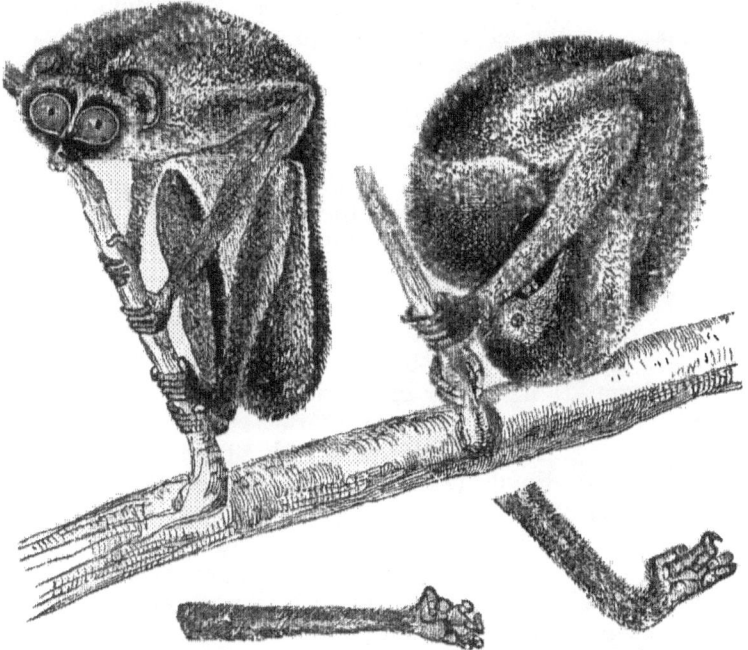

Fig. 32.—The Slender Loris, in waking and sleeping postures, with figures of the arm and leg. (From Sir J. E. Tennent.)*

representing a primitive group of the order (*Primates*) which includes the latter. Lemurs may always be distinguished from monkeys and apes by certain features in their skulls, as well as by several peculiarities in their

* We are indebted to Messrs. Longmans for the use of this figure.

internal structure ; but as these require a certain amount of anatomical knowledge for their proper comprehension, we must ask our readers to take it on trust that such differences do exist. Externally, lemurs differ from ordinary monkeys by their more or less fox-like and immobile countenances, but since the marmosets of South America (which are a lowly type of monkey) resemble them in this respect, this character does not afford an absolute distinction between the two groups. All lemurs are further characterized by having the second toe terminating in a long pointed claw, as shown in the right hand lower figure of our first illustration, whereas in ordinary monkeys the same toe has a flattened nail. Unfortunately, however, the marmosets have also a pointed claw on the toe in question, so that this character does not afford an absolute point of distinction between the two. On examining the upper teeth of the lemurs it will, however, be found that, except in the aye-aye of Madagascar, the first pair of incisor or front teeth are separated from one another in the middle line by a distinct gap, whereas in all monkeys they are, as in ourselves, in contact. Now as the aye-aye differs from all monkeys in having its first upper incisors of the chisel-like form characteristic of rodents (rats, beavers, &c.), the upper front teeth of the lemurs will serve to distinguish them absolutely from the whole of the monkeys.

In appearance the various kinds of lemurs differ greatly from one another, some of them looking not unlike monkeys ; while others, as the one represented in Fig. 32, are characterized by their long and slender limbs, enormous eyes, and general ghostly form. Then, again, while some of them are furnished with long tails, others are destitute of these appendages ; and the common cat-lemur of Madagascar is distinguished from all the rest by the bold alternating rings of black and white with which the tail is ornamented. The last-named species is further peculiar in living chiefly among rocks, whereas the others are arboreal and mainly nocturnal in their habits. It is from these nocturnal habits, coupled with the large eyes, ghostly appearance, and stealthy movements, characteristic of many of the species, that Linnæus was induced to propose the name of lemurs (from the Latin term for evil spirits) for this group of animals—a name which, in the absence of any vernacular title, has been adopted as their ordinary designation.

None of the lemurs are of large size, the length of the head and body in the largest species being only about two feet, and some of them are not larger than rats. They are all excellent climbers, and the majority spend the day sleeping either in the hole of a tree, in a specially constructed nest, or rolled up in a ball after the manner shown in our first illustration. Their food consists of leaves and fruits, birds and the eggs, reptiles and insects, and occasionally honey or sugar-cane; and most of them spend the whole of their time in trees, rarely, if ever, descending to the ground. Some of the larger species inhabiting Madagascar are, however, an exception in this respect, as well as in their diurnal habits, and they may sometimes be observed in numbers jumping across the plains from wood to wood in their own peculiar fashion, when it is necessary to seek fresh food. From the structure of their brains and other parts of their organization, it is evident that the lemurs hold a very low place in the mammalian class, although their near relatives, the monkeys and apes, occupy the highest position. It is probable, indeed, that the modern lemurs are the descendants of the ancient ancestral stock from which monkeys have originated, and since they themselves are also nearly related to the so-called insectivores (shrews, moles, hedgehogs, &c.), while the latter may have been directly descended from marsupials (opossums, &c.), we see how close is the connection between the very highest and the very lowest representatives of the mammalian class. The low position of the lemurs in the zoological scale is in harmony with their antiquity and their peculiar geographical distribution, and it is noteworthy that both in Madagascar and in Africa lemurs are accompanied by certain insectivores of a very low degree of organization, and unlike those found in any other part of the world.

The limits of this essay render our notice of the various

kinds of lemurs necessarily very brief, and our chief
attention will accordingly be directed to some of the more
peculiar and interesting forms. Among the Madagascar
lemurs are included a group known as indris, or, in the
vernacular, sifakas, and containing the largest of all these
animals. These sifakas are distinguished by the circumstance that all the toes of the foot, except the first, are
united together at their bases; and they are further
characterized by their parti-coloured fur, in which white,
black, and various shades of brown and orange predominate. Most of them have long tails, but in one species
this appendage is represented by a mere stump. They
live in small parties in the woods of Madagascar, and feed
entirely upon vegetable substances. By the aid of their
powerful limbs they are able to take enormous flying leaps
—sometimes as much as thirty feet in length—from tree
to tree, and when passing from one clump of trees to
another on the ground they hop on their hind limbs, with
their arms raised above their heads, in a series of short
jumps, when they are said to present the most ludicrous
and grotesque appearance. They are largely diurnal in
their habits, although sleeping during the heat of the
day.

The true lemurs, which are likewise confined to Madagascar and some of the adjacent Comoro islands, differ
from the sifakas in having thirty-six in place of thirty
teeth, by their perfectly free toes, and their less elongated
hind limbs. They all have long tails, and the best known
species is the above-mentioned cat, or ring-tailed lemur,
easily recognized by the feature from which it derives its
second name. This animal, which is often exhibited in
menageries, is about the size of a domestic cat, and is
peculiar in frequenting rocks rather than trees, in at least
certain districts in Madagascar. The black lemur, in which
the male is black and the female red, is a nearly allied
species; and there are also several others. The females of
these lemurs have a peculiar habit of carrying their young
clinging to the under surface of their body, with the head
on one side and the tail on the other; but this strange
position is only maintained for a certain time, after which

the young creature mounts upon its parent's back, where it remains until able to shift for itself.

A third group is formed by the galagos, which differ from the true lemurs in having two of the bones of the ankle greatly elongated, so as to make this segment of the limb much longer than ordinary. They all have long and bushy tails, and are mostly of small size, some of them being even smaller than a rat. The galagos are divided into two groups—one confined to Madagascar and the other to Africa. The former, or dormouse-lemurs, are peculiar in that during the hot dry season several of the species undergo a kind of hibernation, coiled up in the hollow of some tree, and in order to prepare themselves for such a protracted fast they accumulate on their bodies a large store of fat. The true African galagos, in addition to other features, differ from the last in that their large ears are capable of being partially folded up, somewhat after the fashion obtaining in the common long-eared bat. Like the dormouse-lemurs, they are purely nocturnal, and when on the ground they hop after the manner of kangaroos, the elongation of the bones of the ankle doubtless sub-serving this kind of movement.

Widely different from all the above are the curious slow-lemurs of Asia and Africa, of which an Asiatic species is represented in Fig. 32. These lemurs are characterized by the index finger of the hand being either very short or rudimentary, and likewise by their tail being similarly abbreviated. The Asiatic representatives of this group, of which there are two genera and about four species, have the usual three joints to the index finger, which is, however, extremely short, and no trace of a tail. It was one of these lemurs which doubtless suggested to Linnæus his name for the whole group, as their movements are slow and deliberate in the extreme, their eyes large, and their habits completely nocturnal. The Asiatic forms are known by the name of loris. The common loris, with some allied species, extends from Burma through the Malayan region to Siam and Cochin China, and is a solitary animal inhabiting the depths of the forests. The strange and weird little animal represented in our first illustration is

an inhabitant of southern India, and is commonly known
as the slender loris. It differs from the common loris by
its much larger eyes, which are separated from one another
only by a very thin partition, as well as by its slender
body and limbs; and in consequence of these and other
points of difference is referred by naturalists to a distinct
genus, of which it is the sole representative. This beautiful little creature is about the size of a squirrel, and is of
a yellowish-brown colour. It occurs in the forests, usually
in pairs. Sir E. Tennent observes that "the naturally
slow motion of its limbs enables the loris to approach its
prey so stealthily that it seizes birds before they can be
alarmed by its presence. During the day, the one which
I kept was usually asleep in the strange position represented in the figure; its perch firmly grasped with both
hands, its back curved into a ball of soft fur, and its
head hidden deep between its legs. The singularly large
and intense eyes of the loris have attracted the attention
of the Singhalese, who capture the creature for the purpose
of extracting them as charms and love-potions, and they
are said to effect this by holding the little animal to the
fire till the eyeballs burst." I once brought a pair of these
little creatures from Madras to Calcutta, and during the
voyage they lived chiefly on plantains and bread-and-milk.

In West Africa the slow lemurs are represented by the
potto and the smaller awantibo, both of which differ from
the loris by the reduction of the index finger to a mere
stump, and also by the presence of a distinct tail, which
is of greater length in the former than in the latter species.
These animals resemble their Asiatic cousins in their
habits, but are even more deliberate in their movements;
and while the potto appears to be not uncommon, the
awantibo is of extreme rarity.

The whole of the preceding species are included by
naturalists in a single family, but the two following
representatives of the group are so different from all the
others that each is made the type of a distinct family. The
first of these two aberrant creatures is the weird tarsier, of
the forests of Sumatra, Borneo, Celebes, and some of the
Philippine islands. This animal derives its name from

its greatly elongated ankle (*tarsus*), and is rather smaller than an ordinary squirrel, with large ears, enormous eyes, and a long tufted tail. Dr. Guillemard, who when in Celebes was fortunate enough to obtain a living tarsier, writes that "these little creatures, which are arboreal

FIG. 33.—The Tarsier. (From Guillemard's "Cruise of the Marchesa.")*

and of nocturnal habits, are about the size of a small rat, and are covered with remarkably thick woolly fur, which is very short. The tail is long and covered with hair at

---

* Messrs. Murray have kindly lent this figure.

the root and tip, while the middle portion of it is nearly bare. The eyes are enormous, and indeed seem, together with the equally large ears, to constitute the greater part of the face, for the nose and jaw are very small, and the latter is set on, like that of a pug dog, almost at a right angle. The hind limb at once attracts attention from the great length of the tarsal bones, and the hand is equally noticeable for its length, the curious claws with which it is provided, and the extraordinary disc-shaped pulps on the palmar surface of the fingers which probably enable the animal to retain its hold in almost any position. This weird-looking little creature we were unable to keep long in captivity, for we could not get it to eat the cockroaches which were almost the only food with which we could supply it."

Our brief account of the lemurs appropriately closes with the strangest of them all—the now well-known aye-aye of Madagascar. This remarkable animal, whose systematic position was long a puzzle to naturalists, was discovered as far back as 1780, but for eighty years after that was only known in Europe by the single specimen then obtained. The aye-aye differs from all other lemurs in having but eighteen teeth, but its most marked peculiarity is to be found in the circumstance that the single pair of front or incisor teeth in each jaw have chisel-shaped crowns like those of rats and beavers, and grow continuously throughout the life of their owner. Another peculiarity occurs in the extremely long and attenuated third finger of the hand, which surpasses all the others in length, although the whole hand is remarkably elongated. It also differs from other lemurs in that all the toes of the foot, with the exception of the first, have pointed claws, and thus resemble the second toe of the ordinary kinds. In size the aye-aye may be compared to a cat, and it has a rounded and somewhat cat-like head, with a short face, and large naked ears. The tail is long and bushy, and the general colour of the fur dark brown.

The aye-aye represents the extreme development of the lemur type, and it is evident that its peculiarities of structure are correlated with equally well-marked traits of habit. Unfortunately, however, from the nocturnal habits and rarity of the creature itself, as well as from the absence of a sufficient number of competent observers in its native haunts, we are by no means so well acquainted with the habits of this strange creature as is desirable. It appears, however, that the aye-ayes live either in pairs or alone in the bamboo forests of Madagascar, and that they are in the habit of constructing ball-like nests of leaves placed in the forks of trees, to which they retire for their diurnal slumber. The strong incisor teeth are certainly used for ripping up the hard external cases of the sugar-canes, on which these animals delight to feed ; and it is said that they are likewise employed to tear away the bark and wood from the trunks of trees, and thus expose the burrows of wood-eating larvæ, which are then extracted by the aid of the thin middle finger. If any of our readers who have friends living in Madagascar can induce them to try and obtain more particulars in regard to the aye-aye, they will be conferring a real benefit on science.

CHAPTER XI.

ARMADILLOS AND AARD-VARKS.

Of the four animals represented in the figures accompanying the present chapter, three are sufficiently alike to suggest to the ordinary observer their relationship to one another, but the fourth is so utterly different, that it is difficult to point out any important character it has in common with the three others; nevertheless, naturalists generally regard all these four strange creatures as belonging to a single order of mammals, for which the name of Edentata is adopted. The signification of the

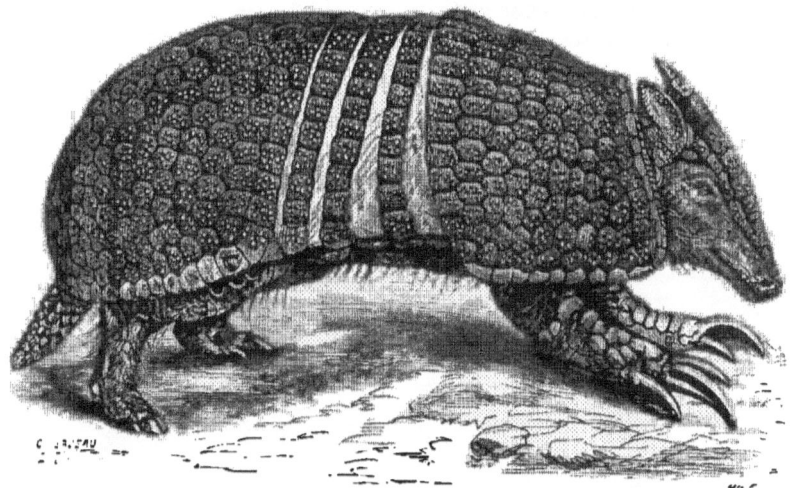

Fig. 34.—The Three-Banded Armadillo.

term Edentata being toothless, the unsophisticated student would naturally be led to suppose that all the animals so named were utterly devoid of those useful but troublesome appendages. This, however, is far from being the case, the majority of the members of the group (among which are those figured here) having a considerable number of teeth.

Still there is one feature in connection with the dentition exhibited by the whole of these so-called edentates; and this is, that teeth in the front of the jaws, corresponding to the incisors of other mammals, are totally absent. Instead, therefore, of being described as edentates, or toothless mammals, these creatures ought rather to have been named aprotodonts, or incisorless mammals. After all, however, it is not much consequence what is the proper meaning of a name, so long as we know the sense in which it is used, and there is accordingly no real objection to the employment of the term edentates, which has obtained the almost universal sanction of zoologists.

In addition to this total absence of front teeth, the edentates are further characterized by the circumstance that all their teeth (when they possess any) show no trace of the hard layer of enamel which is so characteristic and essential a constituent of those of other mammals; these teeth at no period of life forming roots, but continually growing from below. Moreover, in nearly all the edentates there is never any set of milk-teeth developed, although, unfortunately, this cannot be taken as a characteristic of the order, since such teeth occur in one of the armadillos, and also in the animal represented in our fourth figure.

Premising that the edentates are quite distinct from the marsupials and egg-laying mammals, we may say, then, that the only features by which they can be collectively characterized are the want of front teeth, and the absence of enamel on those of the cheek series, while in certain rare instances they may be utterly devoid of teeth. Such characters, it must be confessed, are by no means of first importance.

The mammals thus associated by these negative characteristics, are now chiefly confined to the southern hemisphere, and include the sloths, anteaters, and armadillos of South America, the pangolins, or scaly anteaters of south-eastern Asia and Africa, and the aard-varks of Africa; the true anteaters and pangolins being those in which teeth are wanting. In past times they were also represented by the gigantic megathere, and a number of other allied extinct forms ranging throughout America, which in some respects serve to connect the sloths with the anteaters. This marked restriction of the existing edentates in the southern hemisphere, and their especial abundance in South America, at once stamps them as a very lowly group of animals, there being a well-marked tendency for the preservation of the humbler forms of life in the southern continents and islands of the globe. There is, indeed, a question whether the pangolins, and more especially the aard-varks of the Old World, have any real kinship with the more typical American edentates, but apart from this possibility of the artificial nature of the group as at present constituted there can be no doubt but that all its members are what may be called degraded types—that is to say, that instead of having advanced in the struggle for existence they have lost some of the attributes of the higher animals; evidence of this degradation being afforded by the indications above mentioned of their having been formerly provided with two sets of teeth. In saying that the edentates are lowly and degraded examples of the mammalian type we by no means intend, however, to imply that they are not admirably adapted to their own particular modes of life, or that they have not developed special structures unknown among the higher mammals. In truth, precisely the contrary is the case, since these creatures have taken to modes of life unlike those of the majority of the larger mammals, and have more or less specially modified their structure in accordance with such peculiar habits, so that they are thoroughly and perfectly suited to their environment. And we may add that it is doubtless due to these peculiarities of structure and habits that they have been enabled to survive and hold their own against the competition of the higher forms.

Thus, in the first place, with the exception of a few armadillos, the whole of the edentates are strictly nocturnal; and while the sloths spend the whole of their lives among the branches of the dense forests of South America, some of the others have taken to a burrowing subterranean life. Moreover, the armadillos and the pangolins have

acquired a special protection from their foes in the shape
of a bony or scaly armour, which is a perfectly unique
feature in the whole mammalian class. Another peculiarity
of the group is that no less than three distinct sections of
its members—namely, the anteaters of South America, the
pangolins of southern Asia and Africa, and the African
aard-varks—have taken to feed mainly or exclusively on
termites, or so-called white ants. This practice obviously
gives them an advantage in the struggle of existence, since
with the exception of the marsupial banded anteater and
the egg-laying spiny anteater of Australia (with which,
of course, they do not come into competition), no other
mammals are in the habit of subsisting exclusively on
those insects. And we may notice here that of the three
groups of termite-eating edentates, two—namely, the pangolins and the anteaters—are those which have entirely
lost their teeth ; while in the aard-varks those organs are
retained. As teeth are obviously of no sort of use to
animals subsisting on such a diet, we may regard the two
former groups as those most specially modified for their
particular mode of existence ; and it may thus be suggested
that they have taken to termite-eating for a longer period
than the aard-varks. A similar observation also applies
to the banded and spiny anteaters of the antipodes, the
former retaining a number of minute teeth, while in the
latter they have completely disappeared. Needless to say,
all termite-eating mammals, whether they be edentates,
marsupials, or egg-layers, have extremely long, narrow,
and extensile tongues with which to pick up their insect-
food ; but the presence of such an organ does not, of
course, imply any mutual affinities between the possessors
thereof, and is merely an instance of the similarity of
organs arising from adaptation to a similar mode of
existence. The tongue of the aard-varks is, however, far
less elongated and extensile than that of the pangolins
and true anteaters ; and, therefore, tends to confirm our
suggestion as to the relative duration of time since the
ancestors of these creatures severally took to termite-
eating. Another instance of adaptation displayed by all
the edentates, except the arboreal sloths, is to be found in

the powerful and generally elongated claws or nails with which their feet are armed, such claws being obviously necessary for a fossorial subterranean existence. The aard-varks, as will be seen from our fourth figure, have, however, much shorter and blunter claws than any other member of the group; and this leads me to hazard the suggestion that, in addition to having taken at a comparatively late period to termite-eating, these animals have not been accustomed to a subterranean life for so long a time as their reputed kindred.

Having said thus much as to edentates in general, we must turn to the special consideration of the creatures whose names form the title of this chapter.

The armadillos, as their name (a Spanish one) implies, are distinguished by the solid armour with which their heads and backs are protected; and it is doubtless the peculiar appearance presented by these animals to which we owe the expression " hog-in-armour." In all the armadillo family the armour takes the form of a series of thicker or thinner bony plates imbedded in the skin covering the head and back, and overlain by horny scales; while the under parts of the body and limbs are hairy, and in many species a larger or smaller number of stiff hairs protrude from between the joints of the armour. This bony armour is a perfectly unique feature among existing mammals; and since each plate is ornamented with a more or less elaborate sculptured pattern, such armour when cleaned by maceration forms a most beautiful object. In the true armadillos, as the one represented in Fig. 34, the shield of armour covering the head is quite distinct from that of the body; while the latter is divided into three distinct portions, namely, a large solid shield covering the fore-quarters, and separated by a larger or smaller number of free movable bands occupying the middle of the body from a nearly similar shield protecting the hinder portion of the animal. In our first figured example the number of the movable bands is only three, but they may vary from six to nine (Fig. 36) up to as many as twelve or thirteen in other species. In one extinct armadillo there were, however, no solid shields, the whole body being covered by a series of thirty-two movable bands. The latter species evidently, therefore, leads on to the rare and beautiful little creature represented in Fig. 35, which rejoices in the name of pichiciago. In this tiny animal, which is only about five inches in length, and has a pink-coloured armour above, and long silky white hair below, the armour of the head and body forms a continuous shield of horny plates underlain by very thin plates of bone, and is attached only to the middle line of the back, so that the lateral portions form a kind of cloak loosely overhanging the hairy sides of the body. The hinder end of this cloak is abruptly truncated, and beneath it the hind-quarters of the animal are protected by a solid bony shield, through a notch in the centre of which protrudes

Fig. 35.—The Pichiciago.

the small cylindrical tail. The pichiciago is found on sandy plains only in the western portions of the Argentine pampas. It will be seen from our illustrations that this creature also differs from the true armadillos in the

I

absence of the large external ears which form such a characteristic feature in the physiognomy of the latter.

Reverting to the true armadillos, we find that the majority of the species protect themselves from attack by squatting on the ground, and tucking their limbs within the shelter of the edges of the armour of the body, while the plated head is drawn as close as possible to the front shield. On the other hand, the species represented in Fig. 34 has the power of rolling itself up into a complete ball, like the pill-millipedes of our own country, the wedge-shaped head and tail fitting most perfectly side by side into the deep notches of the front and hind shields. Thus coiled up, the three-banded armadillo is safe from most animals except man. Trusting in this immunity from attack, this armadillo, together with two other species inhabiting the Argentine, has become almost exclusively diurnal in its habits. These diurnal habits, as Mr. W. H. Hudson, in his charming work, "The Naturalist in La Plata," suggests, may also have had the advantage of avoiding any encounters with the larger animals of prey, which are mostly nocturnal, and some of which may have been able to break through the protecting armour, more especially in the species which lack the power of rolling themselves up. Whatever advantage may have formerly accrued from these nocturnal habits before the appearance of man on the scene, is, however, now completely lost in cultivated districts, where these species stand a good chance of being completely exterminated by the hand of man.

On the other hand, the six-banded peludo, or hairy armadillo (Fig. 36), of the Argentine, which differs from its cousins in preferring an omnivorous diet to one of insects, is a far wiser beast in its generation. This creature, according to Mr. Hudson, adapts itself to the conditions under which it exists, and thus stands a good chance of surviving when its fully-armoured relatives perish. "Where nocturnal carnivores are its enemies," writes the observer mentioned, "it is diurnal; but where man appears as a chief persecutor, it becomes nocturnal. It is much hunted for its flesh, dogs being trained for the purpose; yet it

Fig. 36.—Six-Banded Armadillo.

actually becomes more abundant as population increases in any district." As I have witnessed, beneath any decomposing carcase lying in the Argentine pampas the burrow of a peludo is almost sure to be found ; and it is not a little remarkable that the flesh of a creature which has such unpleasant tastes in the matter of diet, should be so eagerly sought after as an article of human consumption.

Before taking leave of the peludo we must not omit to mention two other peculiar habits which are recorded of it by Mr. Hudson, since these also mark it as a creature far above the generality of its kind in point of intelligence. The first of these peculiarities is the ingenious way the creature catches mice, by approaching them with extreme caution, raising itself on its hind-quarters, and then suddenly proceeding to "sit down" on the unfortunate rodents, which become entrapped under the projecting edges of its armour. The sharp edges of the armour are also brought into requisition when this armadillo attacks a snake preparatory to devouring it; the snake being pressed close to the ground beneath the edges of the bony plates, and literally sawn to death by means of a backwards-and-forwards motion of the body of its assailant.

The largest of living armadillos is one which inhabits the moist forests of Brazil and Surinam, and has a length of about thirty-six inches, exclusive of the unusually long tail, which is some twenty inches in length. These dimensions were, however, vastly exceeded by some extinct armadillo-like animals, of which the remains are found in the caverns of Brazil. The most gigantic of these creatures, which flourished during the Pleistocene epoch— the period *par excellence* of giant mammals—is estimated to have been nearly equal in size to a rhinoceros, and has been named the chlamydothere. The armour appears to have been very like that of the true armadillos, but the bony plates measured as much as five and six inches in length, in place of little more than an inch. The teeth differed, however, from the simple conical ones of the modern armadillos, and more nearly resembled the vertically fluted ones characteristic of the extinct glyptodonts. Unfortunately, space does not admit of further reference

to the gigantic creatures from the Pleistocene of South
America, to which the latter name has been applied, all
of which are distinguished from the armadillos by the
armour of the body being welded into a single solid dome-
like shell.

Passing on to the animals whose name comes second in
the title of this chapter, we have first of all to mention
that the designations by which these creatures are commonly known exhibit that remarkable want of originality
in nomenclature which appears to be characteristic of
Europeans when they are brought for the first time into
contact with hitherto unknown animals. Thus, whereas
the Dutch Boers of South Africa applied to the creatures
in question the title of "aard-vark" (meaning "earth-
pig"), the English colonists of the Cape commonly speak
of them as the ant-bear. Now, if there is any one particular animal which the aard-vark (as we must perforce
term the creature) is unlike, it is a bear; while its resemblance to a pig is only of the most distant kind. Still,
however, as in the case of the order to which it belongs,
we must be content to designate the animal by the name
by which it is most commonly known.

In appearance, aard-varks, of which there are two species,
are decidedly ugly creatures, having thick ungainly bodies,
a long pointed snout, enormous erect ears, and a thick
cylindrical and tapering tail, nearly as long as the body.
The skin is either almost naked, or thinly covered with
bristle-like hairs. The fore-feet have but five toes, which
are armed with broad and strong nails, as are the five
toes of the hind limb. As we have already mentioned,
almost the only feature which the aard-vark has in common with the armadillos is the absence of front teeth,
and its cheek-teeth are quite unlike the simple ones of the
latter, as, indeed, they are dissimilar to those of any other
mammals. In the first place, they are preceded by a
functionless series of milk-teeth (a feature found elsewhere among edentates only in one species of armadillo),
while in the second place the premolars are unlike the
molars. The latter are composed of a number of closely
packed denticules, each furnished with a central pulp

FIG. 37.—The Ethiopian Aard-Vark.   (From Sclater, *Proc. Zool. Soc.*)

cavity, and by their close approximation forming polygonal prisms, so that a cross-section of one of these teeth presents the appearance of a pavement. No dental structure among mammals is at all comparable to this, although there is some approximation to it among certain fishes.

Of the two living species of aard-vark, one is confined to South Africa, while the other (represented in our figure) inhabits part of Egypt and other districts in the northwestern portion of the same continent. A third species occurs fossil in the Pliocene deposits of the Isle of Samos, but with this exception the palæontological record is silent as to the past history of these strange creatures, as to whose origin and relationship to the other animals we are at present utterly in the dark. Indeed, the aard-vark is placed among the edentate mammals chiefly because zoologists do not know where else to put it, and they take that group as a kind of refuge for the destitute. Were it not that the burdening of zoological science with new names is from all points of view to be deprecated, there is, indeed, much justification for regarding these animals as the sole representatives of a distinct order; but, although in some ways such a new departure would be convenient, I do not know that in others it would be of any great advantage. But in including them provisionally among the edentates we have to recollect that their affinities with other members of that group—not even excepting the pangolins—must be extremely remote.

Aard-varks lead what would seem to us a very dull and monotonous kind of life, passing the whole of the day curled up in their deep burrows, which are generally excavated hard by the tall pyramidal hills made by the termites, and only issuing forth at night to dig in the mounds for their favourite insect-food. Not a great many years ago it used to be said at the Cape that wherever a clump of termite hills was to be seen there an aard-vark's burrow might be pretty confidently expected. Unfortunately, however, this is no longer the case, and the aard-vark of that district runs a good chance of being exterminated at no very distant date.

This deplorable result is being brought about by the

incessant pursuit of these animals by the natives for the sake of their hides and flesh, and also by their being dug out by Europeans for so-called sport. Their flesh is said to be excellent, and is compared to superior pork; while the value of each hide is about fifteen shillings. This threatened extermination is a very short-sighted policy on the part of the South African farmers, to whom the aard-vark is a valuable ally, not only on account of the enormous number of termites it consumes, but likewise from the circumstance that while it is engaged in digging for these insect-pests it covers with loose earth a quantity of the seeds of grass and other pastoral herbage which would otherwise perish during the hot season. Although there is no likelihood at present of the Ethiopian aard-vark sharing the threatened fate of its southern cousin, yet the extermination of the latter would be a sad loss to zoological science, and we therefore wish every success to a movement which we hear has been set going by the Cape Farmers' Association for the protection of this most strange and curious creature ere it be too late.

Addendum.—Quite recently some remains have been discovered in the upper Eocene, or Oligocene strata of France, which are considered to indicate the existence at that period of small ancestral types of aard-varks.

CHAPTER XII.

THE OLDEST MAMMALS.

Up to the year 1818 it was a generally received axiom of geology that mammals were totally unknown before the Tertiary period; and that period was consequently designated the age of mammals—a name, by the way, which is still perfectly appropriate, if taken to imply that these animals then, and then only, became the dominant inhabitants of the world. In that year, however, the illustrious Cuvier, during a visit to the museum at Oxford, was shown two minute jaws, carrying a number of cusped teeth, which had been obtained in the neighbouring quarries of Stonesfield, from the rock known as the Stonesfield slate, belonging to the lower part of the great Jurassic, or Oolitic, system. After careful examination, the French anatomist pronounced confidently that these two tiny little jaws, neither of which exceeded an inch in length, were those of mammals, and he further suggested that they would prove to belong to a species of opossum. Although this opinion was given in the year 1818, it does not appear that it was published till the year 1825, when the second edition of the fifth volume of the immortal "Ossements Fosiles" saw the light. In publishing this epoch-making notice of the occurrence of mammals in the Secondary period, Cuvier, with the usual caution of naturalists, was careful to add the proviso that everything depended on whether the specimens he saw had really been obtained from the Stonesfield slate. Unfortunately, there does not appear to be any record stating by whom, or at what date, these original specimens—now forming some of the most valued treasures of the Oxford Museum —were obtained from the Stonesfield slate; but that they did come from that formation is perfectly certain. Indeed, other specimens have been subsequently obtained from the same beds, showing certain characteristic Stonesfield shells imbedded in the fragments of rock in which the mammalian jaws are contained. Here we may mention that the Stonesfield slate is the equivalent of the lower portion of the great, or Bath, oolite—a deposit which is separated from the underlying lias by the beds known as the inferior oolite. Conse-

FIG. 38.—One half of the lower Jaw of a Stonesfield Mammal. Twice natural size. The restoration of the front teeth is conjectural.

quently, the mammal-yield-ing beds are separated from the rocks of the Tertiary period, not only by the immense series of Cretaceous deposits (chalk, gault, greensands, and Wealden), but likewise by a large thickness of those belonging to the Jurassic system, such as the Purbeck and Portland oolites, the Kimeridge clay, the coral-rag, and the Oxford clay.

Needless to say, no sooner was the existence of mammals in the Stonesfield slate announced than it was received with a howl of incredulity. First of all it was attempted to show that the specimens themselves did not come from Stonesfield; and no sooner was this objection knocked on the head than doubts were raised as to the Jurassic age of the Stonesfield slate itself. These, however, were equally soon disposed of, and the only thing then remaining was to dispute the mammalian nature of the fossils. This task was undertaken by the French naturalist, De Blainville, who attempted to show that the mammalian character of the specimens was not proved by the double roots of their molar teeth. In the course of the argument, great stress was laid on the circumstance that two-rooted molars were found in a gigantic animal from the Eocene of the United States, then known as *Basilosaurus*, and regarded as a reptile. De Blainville was, however, here treading on very dangerous ground, for it subsequently turned out that *Basilosaurus* itself was really a mammal, which is now generally placed among the whales, under the name of

Zeuglodon. The correctness of Cuvier's original determination was thus in the end triumphantly sustained, and the existence of Jurassic mammals became henceforth an established fact in geology, although the suggestion that these fossils belonged to opossums was, of course, unfounded.

Passing on to the consideration of the specimens themselves, we find that the great peculiarity of the jaws of these Stonesfield mammals (for one of which De Blainville proposed the name of amphithere) is the excessive number of their cheek-teeth, a feature now paralleled only in the little banded anteater of Australia. This multiplicity of teeth is well shown in the jaw represented in Fig. 38, which is preserved in the museum at York, and shows upwards of nine cheek-teeth still remaining, whereas in practically all existing mammals with complex teeth, except the banded anteater, the number does not exceed seven. Other jaws were, however, subsequently discovered in Stonesfield, in which the number of cheek-teeth was considerably less; but one of these later specimens (described as the phascolothere) revealed the important fact that there were four pairs of front or incisor teeth in the lower jaw. Now since it is only among the pouched mammals, or marsupials, that more than three pairs of incisor teeth are found, while the banded anteater, with its numerous cheek-teeth, is a member of the same group, it became a very natural conclusion that the Stonesfield mammals were likewise marsupials. Support was lent to this conclusion by the circumstance that, with the exception of the egg-laying mammals, or monotremes, the marsupials are the lowest of all living mammals. And indirectly some further support to this view is afforded by the fact that Australia still retains other forms of animal life allied to those which were living in Europe during the period of the Stonesfield slate. For instance, it is in the Australian seas alone that there still survives the solitary representative of the beautiful genus of bivalve shells known as *Trigonia*, which were so especially abundant in the oolites; while it is also there alone that swims the Port Jackson shark, whose mouth is armed with a pavement of crushing teeth, recalling those of many of its Jurassic forerunners. Moreover, in the presence of numerous cycads among its flora, Australia again recalls the Jurassic epoch of Europe; and it has accordingly been suggested that modern Australia may be regarded as a kind of direct survival from Jurassic times. Before, however, we can say anything more as to the affinities of the Stonesfield mammals, we must turn our attention to subsequent discoveries of mammalian remains in other formations.

The first of these discoveries was made in the year 1847, by Professor Plieninger, of Stuttgart, who obtained certain minute teeth from a bone-bed near that city belonging to the upper part of the Triassic period, which were declared to be mammalian, and for the owner of which the name *Microlestes* was proposed. Now, as the trias lies below the lias, the existence of mammalian life was by this discovery carried back at one bound very nearly to the commencement of the Secondary period. Subsequently, mammalian remains were obtained from the trias of Somerset, which proved to belong to the same genus as those from Stuttgart, while others of a different type were found in the equivalent deposits of North America.

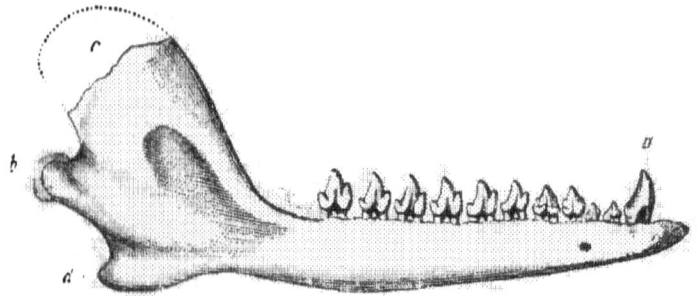

Fig. 39.—Lower Jaw of an American Jurassic Mammal; twice natural size. (After Marsh.)

A little later, the year 1854 was made memorable by the first discovery of mammalian remains in the freshwater Purbeck strata of Dorsetshire, belonging to the very top of the Jurassic system; from which formation in subsequent

years a vast number of such remains were obtained, through
the energy of the Rev. P. B. Brodie and Mr. Beccles. All
these specimens were obtained from a single bed, and
many of them indicated forms more or less closely allied
to those from Stonesfield and Stuttgart. It was thus
shown, once for all, that mammalian life must have been
locally abundant throughout the Jurassic period. This
conclusion was subsequently amplified by the discovery in
the upper Jurassic rocks of North America of a whole

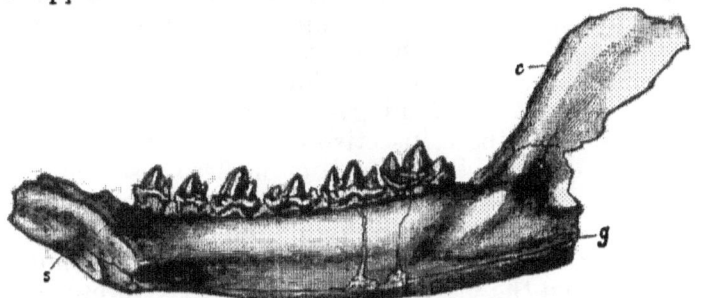

FIG. 40.—Lower Jaw of *Triconodon* ; twice natural size.
(After Marsh.)

host of small mammals very closely allied to those from
Dorsetshire, a large number of which have been described
by Prof. O. C. Marsh, some of whose figures are here
reproduced. Many of these small Jurassic mammals
(Figs. 39 and 40) were evidently carnivorous, and such
carnivorous forms exhibited two distinct types of dentition.
In one of them (Fig. 39) there was a numerous series of
cheek-teeth behind the tusk, or canine (*a*), each of which
carried three cusps arranged in a triangle; while in the
other type (Fig. 40) the cheek-teeth were fewer in number,
and had the three cusps on their crowns ranged in the same
line. From this peculiarity the animal to which the second
type of jaw belonged was appropriately named *Triconodon*.
The first type corresponds to the amphithere of the Stonesfield slate, while the second is more like the phascolothere
of the same formation.

In Fig. 40 it will be observed that there is a peculiar groove (*g*) running along the inside of the jaw, and

since a similar groove is found among existing mammals only in the banded anteater and certain other carnivorous marsupials, we have pretty conclusive evidence that *Tricon-odon* and its allies were really marsupials. There can also be but little doubt that the species of the amphitherian type (I avoid mentioning the numerous genera of these animals) are likewise members of the same order. It is, however, quite possible that some of the Jurassic mammals of Dorsetshire and North America may be more nearly allied to that primitive group of mammals known as Insectivores, among which are included the mole, the shrew, and the hedgehog of Europe, as well as the more generalized tenrec of Madagascar, and many other peculiar creatures. All these insectivores are of a very low grade of organization, and the result of modern researches is to show that their connection with the marsupials is very close indeed. Hence it is highly likely that some of the Jurassic mammals may have been the actual connecting links between the marsupials and the insectivores; and it is worthy of mention here that while marsupials at the present day linger on only in Australia and America, some of the most primitive types of insectivores are preserved to us in Madagascar, which is another refuge for animals of a low grade of organization.

There is yet another type of mammal found in the English and American Jurassics, to which the *Micro-lestes* of the trias also appears to belong, which has given rise to a vast amount

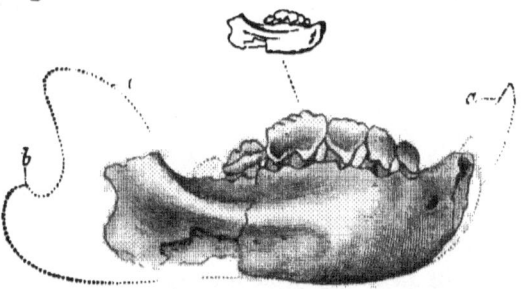

Fig. 41.—Lower Jaw of *Plagiaulax;* natural size and enlarged. (After Marsh.)

of discussion among the palæontologists. These remarkable mammals are mostly of very minute size, and were long known only by their lower jaws, of which a specimen is represented in Fig. 41, from which it will

be seen at a glance that the dentition is quite different
from that of either of the carnivorous types figured above.
The lower teeth comprise a single large incisor (*a*), behind
which were either three or four tall premolar teeth with
cutting edges, and marked on the sides with a number of
oblique grooves, from which the name *Plagiaulax* was
taken. When unworn, these grooves extended along the
whole outer surface of the teeth, but when the teeth had
been long in use (as in our figure) the groovings became
worn away from the sides. Behind these four premolars
are two smaller molar teeth, with the summits of their
crowns marked by a single longitudinal groove bounded
by prominent ridges. Now it was argued at first that
this very peculiar type of dentition indicated carnivorous
habits in the owners thereof; but it was subsequently
pointed out that the existing rat-kangaroos of Australia
(of which the front of the skull is shown in Fig. 42)
presented a some-
what similar type
of tooth-structure.
Thus, the last pre-
molar tooth (*p.m.*)
of the rat-kanga-
roo has a cutting-
crown marked
with a number of
parallel grooves;
while each half
of the lower jaw
terminates in a

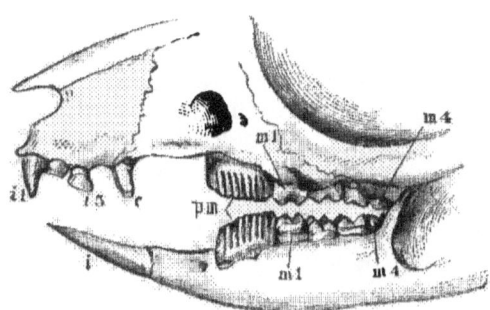

Fig. 42.—Jaws and Teeth of Rat-Kangaroo.

single large incisor not unlike that of the Jurassic *Plagiaulax*. Hence it was argued—and, in our opinion, argued
rightly—that as the living form is herbivorous, the same
must have been the case with the extinct one. When,
however, it was also urged that the rat-kangaroo and
Plagiaulax were closely allied animals, important differences
between the two were overlooked. For instance, as will
be apparent from the figures, while in the former there
was but one grooved tooth, in which the grooves are
vertical, in the latter there were usually three or four

such teeth in which the grooves are *oblique*. Moreover, whereas the recent form was provided with four molar teeth (*m* 1—*m* 4), the fossil had but two such teeth; while the form of these teeth was quite unlike in the two. Hence, when we add that there are other important differences between them (into the consideration of which it would be difficult to enter here), it will be apparent that the view of regarding *Plagiaulax* as a near ally of the rat-kangaroo was the result of attaching too much importance to resemblances, and overlooking differences. Indeed, such resemblances as do exist between the two may be regarded merely as a well-marked instance of the phenomenon of parallelism, treated of in an earlier chapter.

Taking it, then, as proved that *Plagiaulax* is not a near ally of the rat-kangaroo, we have to consider whether it can be affiliated to any other group of existing mammals. Before doing so, we have, however, to mention that there are certain other Secondary mammals allied to *Plagiaulax*, in which the whole of the cheek-teeth are like the true molars of the latter. We have already stated that in *Plagiaulax* the lower molars have a median longitudinal groove, and it may be added that the ridges bordering such grooves are surmounted by a number of small tubercles. In the upper jaw, if we may judge by some allied genera, the molars had three such tubercular ridges, separated by two grooves. Similar molars occur in the skull represented in Fig. 43, which is that of a mammal discovered a few years ago in the Secondary rocks of South Africa, and named by Sir R. Owen *Tritylodon*; but it will be noticed that there is no trace of the cutting and obliquely-grooved premolar teeth of *Plagiaulax*, the premolars being like the molars. Detached molars of similar type have been found in the Trias of Stuttgart, and others occur in the Stonesfield slate. Moreover, in Dorsetshire and North America, there are certain nearly allied mammals (*Bolodon*) in which the upper molars have only two, in place of three, longitudinal ridges of tubercles. These forms, if other proofs were wanting, clearly show, indeed, that the resemblance between *Plagiaulax* and the rat-kangaroo is not a genetic one. When, however, the molar teeth of the type in which there are but two longitudinal rows of tubercles are compared with the transitory teeth of the Australian duck-bill, a certain resemblance can be detected between the two, which seems sufficient to indicate that in *Plagiaulax* and *Tritylodon* we have to do in all probability with ancient types of egg-laying mammals.

Till within the last few years the Cretaceous period formed a complete gap as regards the history of mammals; and seeing that in Europe, with the exception of the Wealden, the rocks of this system are mainly of marine origin, while some of them, like the chalk, were laid down in seas of considerable depth, this absence of mammalian remains is not to be wondered at. In the United States the condition of things is, however, very different. There the upper-most Cretaceous rocks are of fresh-water origin, and constitute a series known as the Laramie, which is in

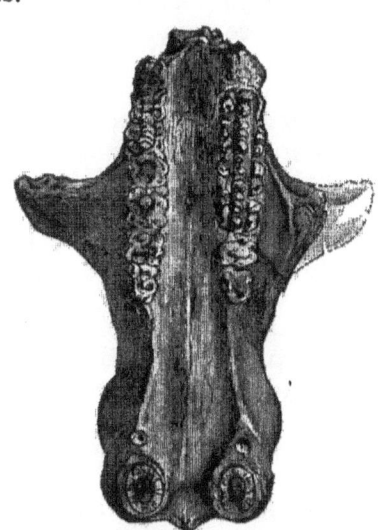

FIG. 43. — Under part of the Skull of a South African Secondary Mammal. Two-thirds natural size.

intimate connection with the lower part of the Tertiary, and has yielded the extraordinary horned dinosaurs. From these Laramie Cretaceous rocks Prof. Marsh has succeeded in obtaining a quantity of teeth of mammals, although these are, unfortunately, mostly found detached. These teeth indicate mammals closely allied to *Plagiaulax* of the Jurassic, and also others of a carnivorous type related to *Amphitherium*, or some of the many-molared carnivorous forms from the Dorsetshire Purbeck. Mammals of the *Triconodont* type—that is, those with the three cusps of the molars in a straight line—seem, however, by this time

K

to have totally disappeared.   At a still later date a single
tooth of the *Plagiaulax* type, as well as one allied to
Bolodon, have been described from the English Wealden,
indicating that at that epoch the Purbeck mammals still
survived in Europe, and leading to the hope that future
researches will yield us further evidence of the European
mammalian fauna of the Wealden.

The present state of our knowledge, therefore, shows
that from nearly the lowest beds of the Secondary period
till the close of that vast epoch, there existed a numerous
fauna of small mammals distributed over a large portion
of the globe, and displaying a remarkable persistence of
nearly similar type.   It is further evident that such of
these mammals as exhibit a carnivorous type of dentition
appear to be allied to the more primitive of the existing
marsupials, although some of them may be more nearly
related to the almost equally low insectivores.   On the
other hand, those which exhibit what appears to be an
herbivorous modification of dental structure, if they are
related to any living forms, appear to have an affinity with
the modern egg-laying mammals of Australia.   Now the
latter, together with the marsupials and insectivores, being
the lowest representatives of mammalian life at present
existing, are precisely such mammals as we should naturally have expected to have been foreshadowed by more
or less nearly allied forms in the Secondary rocks ; and,
therefore, in this respect, theoretical palæontology is, so
far as our present knowledge goes, precisely in accord
with actual facts.

That the few Triassic mammals at present known were
the earliest representatives of the class cannot, however,
be admitted for a moment, and we must accordingly look
either to the lower Triassic rocks, or to those of the
underlying Permian (forming the top of the Palæozoic
series), for the discovery of such primitive types.   Should
such ever be discovered, it is to be confidently expected
that they will exhibit such a combination of characters
common to mammals, and certain extinct reptiles and
amphibians, that it will be very hard to say under what
class they will have to be ranked.

Seeing that throughout the whole of the Secondary period, with the possible exception of a few lowly insectivores, there is no evidence of the existence of any mammals belonging to the higher placental type (under which are included all living representatives of the class save the marsupial and egg-laying groups), the reader will naturally inquire when such higher forms first made their appearance. We answer, with the first dawn of the Tertiary period; for in the very lowest Eocene strata both of France and the United States there are found, side by side with small mammals allied to *Plagiaulax* and the marsupials of the Jurassic and Cretaceous, others, which, though still of small size, were evidently placentals. And it is very remarkable that this first definite appearance of the higher forms of mammalian life should, so far as we know, have been contemporaneous with the disappearance of so many gigantic types of extinct reptiles, such as the dinosaurs, the fish-lizards, and the plesiosaurs, which seem to have reached the end of their term of existence at or about the close of the Secondary period.

Most of these early Tertiary mammals had molar teeth carrying three cusps arranged in a triangle, like their marsupial forerunners of the Secondary epoch, from which, indeed, they were probably derived; and at this comparatively early epoch the orders of mammals were but very imperfectly differentiated from one another, it being frequently difficult to decide which were carnivores and which were ungulates. A few stages later differentiation of ordinal types, accompanied by a great increase in the bodily size of their representatives, had, however, taken place; and by the close of the Eocene period, as exemplified by the higher deposits of the Paris basin, most of the present orders of mammals were well defined. Thence, through the succeeding Miocene and Pliocene epochs, there went on a continual evolution of mammalian life, resulting in the production of giant forms like the elephant and the rhinoceros, and also characterized by the development of specially modified types like the horse and the ox, which differ so widely from their five-toed ancestors. During the same epochs antlers were developed in the

K 2

deer and horns in the rhinoceroses and oxen, while pigs and hippopotami gradually acquired the enormous tusks with which their existing representatives are armed.

Seeing, then, that it was not till the advent of the Tertiary period that mammals assumed the position of the dominant forms of life, Cuvier's memorable discovery in the second decade of this century that the class dated from the middle of the Secondary period has in no essential respect served to dispossess the first-named epoch from its claim to the title of the AGE OF MAMMALS.

CHAPTER XIII.

CROCODILES AND ALLIGATORS.

IN spite of the circumstance that numerous examples of those ungainly reptiles known as crocodiles and alligators are exhibited in the reptile house of the Zoological Society's Gardens in a living condition, while their stuffed skins and articulated skeletons are displayed in the galleries of the Natural History Museum at South Kensington, there appears to be a hopeless confusion in the public mind between these two very different creatures. And, with the usual perversity of those not acquainted with the ordinary facts of natural history, residents in India increase this confusion by almost invariably speaking of the crocodiles of that country as alligators, whereas an alligator is not to be found from one end of India to another. A remarkable instance of this confusion occurs in Sir S. Baker's " Wild Beasts and their Ways," where, under the heading of crocodile, it is stated that, "as lizards are found distributed in great varieties throughout the world, in like manner we find the largest of all lizards, the crocodile, under various names in nearly every river of the tropics. In America this reptile is generally known as an alligator, and some persons pretend to define the peculiarity which distinguishes that variety from the crocodile, but I regard the distinction in the same light as that between the leopard and the panther, the difference existing merely in a name."

Now, in the first place, although it may be justifiable in popular language to use the term "lizard" as applicable to all four-footed reptiles except tortoises and turtles, yet, scientifically speaking, a crocodile has not the slightest right to be so termed. Indeed, it would be far preferable to speak of a snake as a kind of lizard, since it is really only a special modification of the lizard stock; and from a strictly scientific point of view it would imply much less

confusion of ideas to call a cow a kind of pig than to term
a crocodile a lizard, since whereas a cow and a pig are
mammals belonging to the same section of a single order,
lizards and crocodiles represent two totally distinct orders
of reptiles.   With regard to the statement that the difference between a crocodile and an alligator is merely one
of name, the reader who follows us through this chapter
will probably hold a different opinion by the time he
reaches the end.

So far as external appearance goes, most people are
aware that crocodiles and alligators are large, long-tailed,
low-bodied reptiles, with flat and frequently broad heads,
and their bodies protected by a coat of scales, which vary
greatly in size in its different regions.   They probably also
know that it is the impressions of these scales, or of the
bony scutes by which those of the back are underlain, that
form the well-known markings on the crocodile-skin now
so commonly used for bags and other leather articles.   In
their short and clawed limbs there are five toes in the
front pair, and four in the hinder, those of the latter being
connected together for a part of their length by a web.
As regards their habits, crocodiles and alligators are typical
amphibious creatures, being perfectly at home in the
water, but also capable of active progress on land, on which
their eggs are laid and the young hatched.   The position
of their external nostrils at the very tip of the snout
enables them to come to the surface for the purpose of
breathing without showing more than their muzzle, or, at
most, this and their somewhat prominent eyes.   These
external characters will enable us to recognize an alligator
or crocodile when we see it, and yet do not show us how
these creatures differ so essentially from true lizards as to
render it incorrect to speak of them merely as a particular
group of lizards.   To render this essential distinction
apparent we must enter into certain details of their
anatomical structure, more especially as regards the skull.
Now, in the first place, a crocodile or alligator may be
at once distinguished from every true lizard by the
circumstance that its large and pointed teeth are inserted
in the jaws in distinct and separate sockets, from which

they will readily fall out in a dried skull; whereas those of lizards, which vary greatly in form, are invariably united by solid bone with the edges or sides of the jaws, without any separate socket. Moreover, in a crocodile's skull, there is a bar of bone running backwards from the lower border of the eye-socket, or orbit (Fig. 44, *O*), to join the condyle with which the lower jaw articulates. This bar is seen in Fig. 44, below,

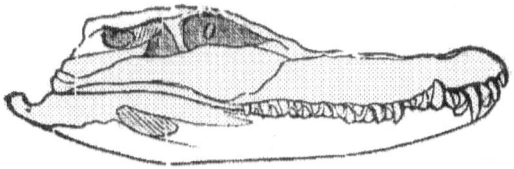

and to the left of the letter *O*, and also occupying the same relative position in Fig. 46. It will further

Fig. 44.—Side view of the Skull of a Crocodile; *O*, eye-socket or orbit.

be apparent from the latter figure that in a crocodile's skull there are two parallel bars running backwards from behind the orbit, of which the upper one is the stouter. Now in a lizard's skull, only the uppermost of these two bars is present; and we thus have a second important distinction between a crocodile and a lizard. A still more important difference occurs, however, in the under part of the skull. Thus, whereas in a lizard the external nostrils open directly through the palate into the front part of the mouth, in a crocodile the bones of the palate develop a kind of flooring beneath its roof, and thus form a closed passage by which the internal or posterior nostrils are brought to the very hinder extremity of the skull; this remarkable peculiarity being well exhibited in Fig. 45, A, where *N* indicates the internal nostrils. The small round aperture seen in the front of the palate in both A and B is closed during life with membrane, and thus prevents any communication between the external nostrils and the front of the mouth. The object of this peculiar arrangement is to enable the animal to breathe when its mouth is open under water, and the nostrils are alone in the air; this being effected by the closing of the back of the mouth, in front of the internal nostrils, by means of a fold of skin, thus leaving a free communication between the nostrils and the wind-pipe. The advantage of this arrangement to animals which, like crocodiles and alligators, kill their prey by holding them under water, is self-apparent.

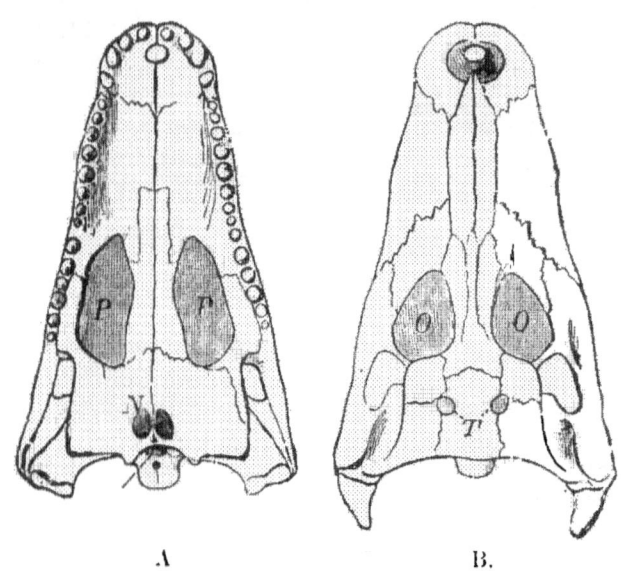

FIG. 45.—Lower view (A) of Skull without the lower jaw, and (B) upper view of Skull with lower jaw of a Crocodile. *O*, orbit; *T*, temporal pit; *P*, palatal vacuity; *N*, internal nostrils.

If to these differences between the skulls and teeth of crocodiles and lizards, we add that, whereas in the latter the ribs articulate to the joints of the back-bone or vertebræ by means of little knobs on the sides of the vertebræ themselves, in crocodiles they join the summits of long horizontal processes of bone projecting from the upper part of these vertebræ, we think we shall have said enough to convince our readers that it is altogether incorrect to speak of crocodiles and alligators as lizards. Crocodiles are, indeed, first cousins of those extinct reptiles known as dinosaurs, and ought, therefore, to be regarded with respect as being the sole existing, although collateral, representatives of that great group of reptiles which dominated the earth at a time when mammals were only just beginning their career.

Having said thus much as to the distinctness of crocodiles from lizards, we may proceed to consider how the former differ from alligators, and to make some mention of a few of the various species of each. Now if we examine the skulls of the different kinds of crocodiles we shall find that the number of teeth in the upper jaw varies from seventeen to nineteen, while in the lower jaw there are invariably fifteen ; and we shall likewise find that the teeth of the two jaws interlock with one another when the mouth is closed. Moreover, when the jaws are in apposition it will be observed that the first tooth on each side of the lower jaw is received into a pit in the palate of the skull, while the fourth lower tooth, which (like the first) is larger than the others, bites into a notch in the side of the skull (as shown in Fig. 44), and is thus more or less distinctly visible externally in the living animal. Crocodiles are now found in the rivers of Africa, India, Burma, Australia, and America, as well as in many of the larger islands in warm regions. They vary greatly in regard to the relative length of the skull, the longest-snouted species occurring in South America and West Africa, while those of India have the shortest and broadest skulls (Fig. 45). On the other hand, if we examine the skull of an alligator, we shall find that the upper teeth bite on the outer side of the lower ones without any sort of interlocking, and both the first and the fourth lower teeth are received into pits in the skull, so that when the mouth is closed both of them are totally invisible from the outer side. Moreover, in no alligator does the number of lower teeth ever fall short of seventeen. Again, in all alligators the skull is even broader and shorter than in the Indian crocodiles. Till within the last few years it was believed (in spite of the persistent assertion of sportsmen, that the Indian "magars," as they are called by the natives, are alligators)

that alligators were confined to the New World, but
recently it has been found that one species exists in
China. This is, indeed, a very curious instance of what
is known as discontinuous distribution, and one which
only finds a complete parallel in the case of the tapirs, of
which there is one species inhabiting the Malay peninsula
and adjacent islands, while all the others are restricted to
South America. There are several species of alligators,
which are divided into two groups, according as to whether
an armour of bony plates, or scutes, is or is not developed
on the under surface of the body. In the true alligators,
which agree with all living crocodilians in having a dorsal
armour of these bony scutes, the upper teeth vary from
seventeen to twenty in number, and those of the lower
from eighteen to twenty, while there is no bony armour
on the under surface of the body. The two well-known
species are the Mississippi and the Chinese alligator, in
addition to which there is a third American form of
which the exact habitat is unknown. The second group
of alligators, or caimans, as they are called in Brazil, is
confined to South America, where it is represented by five
species. These are characterized by having from eighteen
to twenty upper, and from seventeen to twenty-two lower
teeth on each side, and also by the lower surface of the
body being protected by a shield of bony scutes, which
overlap one another like the tiles on a roof, and each of
which is composed of two separate pieces united together
by what is known as a sutural union.

The above, then, are the chief differences which distinguish alligators and caimans on the one hand from
crocodiles on the other, and they are such as surely do not
justify the statement that naturalists merely pretend to
distinguish between the two. Alligators and crocodiles do
not, however, exhaust the list of living crocodilians, since
we have two peculiar species differing from all the others
by the great length of their snouts, and respectively
inhabiting the Ganges and the rivers of Borneo—the
former being known as the garial, and the latter as
Schlegel's garial. In both of these reptiles the numerous
teeth are long and slender, and differ from one another

but little in size in different parts of the jaw, while
neither of them have an armour on the lower surface of
the body. They differ from crocodiles and alligators in
feeding chiefly on fish.

In regard to their geological distribution, crocodilians
more or less closely allied to the existing garials, crocodiles,
and alligators, are found throughout the rocks of the
Tertiary period as far down as the London clay. Some of
these extinct species were, however, of gigantic dimensions;
one from the Siwalik Hills of India, which was allied to
the garial, attaining a length of between fifty and sixty
feet, and thus presenting a great contrast to living crocodiles, which rarely exceed a length of some twenty-two or
twenty-three feet. Another species found in the Tertiary
clays of Hampshire presents characters intermediate
between crocodiles and alligators, the fourth lower tooth
being usually received into a pit in the skull, but the
under surface of the body having a complete bony armour
like that of the caimans. This genus (*Diplocynodon*)
differs from both crocodiles and alligators in that both
the third and fourth lower teeth are larger than the
adjacent ones, so that the animal had two powerful tusks
in the sides of the lower jaw.

A few crocodilians, more or less closely allied to existing
types, also occur in the Cretaceous rocks, but when we
reach the Wealden and Jurassic strata nearly all the forms
differ very markedly from modern ones, and show a lower
stage of development. Before, however, we are in a
position to understand how these early crocodilians differ
from their living cousins, we must enter a little more fully
into the anatomy of the latter. Turning once more to
Fig. 45, we see that the passage leading to the internal
nostrils is formed by four pairs of bones, on the fourth of
which the letter N is placed. Again, in all living crocodilians the vertebræ articulate with one another by a ball-
and-socket joint, of which the cup is situated at the front
of each vertebra; this mode of articulation being the
best adapted to give free motion of one vertebra upon the
other. The third point we have to notice relates to the
bony armour of living crocodilians, in all of which the

pitted scutes forming the shield on the back are ridged,
and arranged in from four to eight longitudinal rows.
Moreover, in the caimans and the extinct *Diplocynodon*, the
shield on the under surface of the body forms a single
mass, made up of more than eight longitudinal rows of
scutes, each of which, as already mentioned, consists of
two separate pieces united together by suture.

If we now contrast these features with those obtaining in
the Jurassic crocodilians, we shall find very considerable
differences. Thus in the skull of those reptiles the fourth
pair of bones on the palate did not meet in the middle line
below the passage to the nostrils, so that the internal nostrils were placed immediately behind, or sometimes partly
between, the bones lying between *PP* in Fig. 45, and were
thus much forwarder than in modern crocodilians. Then,
again, the vertebræ were slightly cupped at both ends, thus
admitting of much less motion between one another. The
armour on the back of the body is of a similar type, consisting only of two longitudinal rows of scutes, which lack the
longitudinal ridges so characteristic of those of the existing
forms. On the other hand, the armour on the under surface
was nearly always present and more developed, frequently
consisting of two distinct portions, in the foremost of which
the scutes (which consisted of a single piece) overlapped
like slates, while in the hinder part they were joined by
their edges to form a solid pavement of bone.

Like their modern cousins, the Secondary crocodilians
included both long-snouted (Fig. 46) and short-snouted

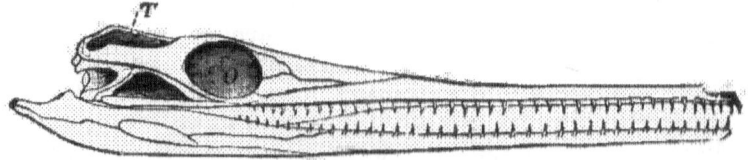

FIG. 46.—Side view of the Skull of an Extinct Crocodilian of the
Lias; one-fourth natural size. Letters as in Fig. 45.

types, the former being the more common and especially
abundant in the lias, where their remains occur in company with those of fish-lizards and plesiosaurs. In many

of these forms the pit (T) in the temporal region of the skull was larger than the socket of the eye, whereas in recent crocodiles it is much smaller (Fig. 45), and in the alligators may even disappear. A large number of these Jurassic forms were of marine habits, and a few of them attained enormous dimensions, the length of the skull of one species falling not much short of five feet. A few species are further peculiar in having altogether discarded their bony armour on both surfaces of the body.

Looking at crocodilians as a whole, it is perfectly evident that they have advanced in complexity of organization with the progress of time, the backwardly-placed internal nostrils and ball-and-socket vertebræ of the modern types being clearly an advance on the Jurassic forms. As, however, is the case in many similar instances, the gradual backward shifting of the internal nostrils presents a problem difficult to understand, as it is hard to conceive what advantage the species in which these nostrils were situated in the middle of the palate had gained over reptiles in which they were placed near the muzzle, the completely backward position being apparently essential in order that the mouth might be kept open under water.

With regard to the general disappearance of the inferior body-armour and the invariably increased development of that on the back of the recent forms, it may be suggested that, as most of the Jurassic crocodilians were of marine habits, and probably swam far out to sea, it would have been highly advantageous for them to have the lower surface of the body protected from attacks from below, by sharks and other creatures. On the other hand, since modern crocodiles and alligators spend a considerable portion of their time on the banks of rivers, and when in the water are in the habit of reposing or crawling on the bottom, it is obvious that the back is the portion which requires especial protection. An explanation of the existence of an armour on the lower surface of the body in the caimans is less easy to give, although it may be merely an instance of the retention of an ancestral character.

We may conclude this account by referring to some interesting observations on the eggs and embryos of the

crocodile of the Nile. It appears that in Madagascar
the egg-laying lasts from the end of August to the end of
September, the number of eggs in a nest varying from
twenty to thirty. The nest is dug about two feet deep in
the dry white sand ; the bases of its wall are gouged out,
and into the lateral excavations thus formed the eggs roll
from the slightly raised centre of the floor of the nest.
Externally the nest is not discernible, but the parent
sleeps upon it. The eggs differ greatly in form ; the shell
is white, thick, firm, and either rough or smooth, the
double shell-membrane being so strong that the egg keeps
its form after the shell has been removed. When newly
laid, the eggs are very sensitive, and are readily killed by
damp or by heat, but the older eggs are hardy. When
the young embryos are about to be hatched, they utter
distinct notes, which the mother hears, even through two
feet of sand, and proceeds to dig open the nest. Before
hatching the embryo turns, and in so doing partially tears
the foetal membranes. With the tip of its snout turned
to one end of the egg, the young animal bores through
the shell with a double-pointed tooth comparable to that
which young birds possess. This tooth appears very
early—by the time the embryo is six weeks or two months
old—and may still be seen a fortnight after hatching.
Through the small perforation made by this tooth the
fluid flows out, softening the adjacent parts, so that the
aperture is widened into a cleft. The process of creeping
out may take about two hours. The young animal seems
large in comparison with the egg ; one measuring 28 cm.
in length came out of an egg 8 cm. long and 5 cm. broad.
The young crocodiles are wild little animals, and are led
to the water by the mother. They utter sounds, especially
when hungry, but the pitch of their call is not so high as
it was when they were within the egg.

CHAPTER XIV.

THE OLDEST FISHES AND THEIR FINS.

IT is a well-known fact that while the fishes of the more
ancient periods of the earth's history were frequently
characterized by having their bodies protected by a coat of
armour, this armour has been lost by most of their modern
descendants.   Nevertheless, a few of these mail-clad fishes,
like the gar-pike of the rivers of North America, and the
many-finned bichir (*Polypterus*) of the upper Nile and the

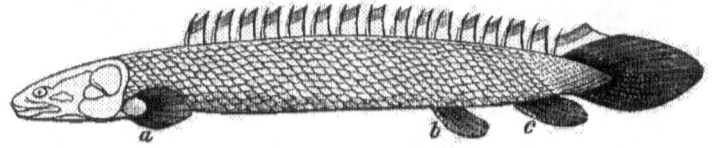

FIG. 47.—The Bichir.

rivers of the West Coast, still linger on, as if for the
purpose of showing us what their ancestors were like.

In addition, however, to their armour, and its gradual
loss with the advance of time, there are many other points
of view from which these ancient fishes are of more than
ordinary interest, and we accordingly propose in this
chapter to consider the curious modifications which have
taken place in the structure of their fins as we ascend in
the geological scale.   We shall, moreover, be led to notice
briefly one of the most remarkable types of fossil fish
teeth found in the older Secondary rocks, since we can
thereby prove that one of our living fishes is the oldest
kind of vertebrate now inhabiting the earth.

Before going further, we must mention that existing
fishes have been divided into several main groups, distinguished from one another by structural peculiarities.
One such group includes the sharks and rays, characterized
by their cell-like gills and scaleless bodies.   Then we have
the smaller group of lung-fishes, now represented by the

lung-fish of Queensland, and the mud-fishes of the rivers of Africa and South America, all of which can breathe either by gills or by lungs. Another group is formed by the so-called ganoid fishes (Fig. 48), many of which have the bony armour already mentioned; while the great majority of the fishes of the present day, although nearly related to these ancient ganoids, have been generally separated as a distinct group, under the title of bony fishes. That name they take from the circumstance that their skeletons are fully ossified, and do not partake of the cartilaginous nature of those of a shark or a ganoid.

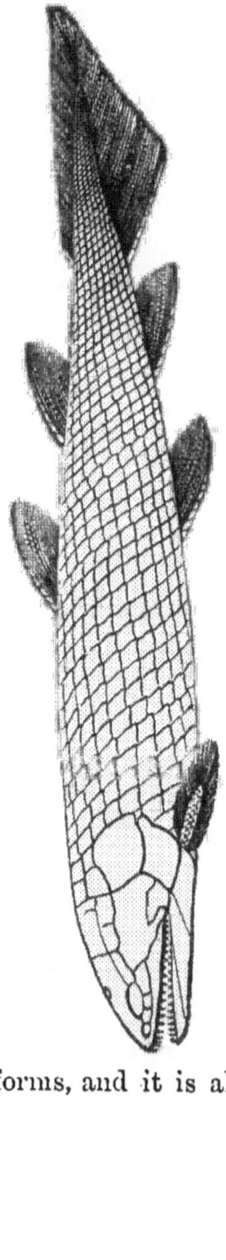

Now if we look at the paired fins (or those which correspond with our own limbs) of any ordinary bony fish, such as the perch (Fig. 49), we shall see that they are formed of a number of bony rays, starting from a single point of origin, and thence spreading out in a fan-like manner. We shall also not fail to observe that the tail of such a fish has a very similar kind of structure, likewise consisting of bony rays, symmetrically arranged, and starting from a curved line where the scales suddenly stop. We may also see from the figure of the skeleton of such a fish that the backbone likewise stops suddenly where the tail begins, and that the rays of the latter start from the expanded end of the backbone itself.

The same general type of structure obtains in the paired fins of some of the later ganoids, such as the living sturgeons, as well as certain extinct forms, and it is also

FIG. 48.—An Extinct Ganoid Fish.

universally present in all the living sharks and rays. If, however, we were to go to the Natural History Museum and examine the lung-fish of the rivers of Queensland, or the gar-pike (*Lepidosteus*) of those of North America, or

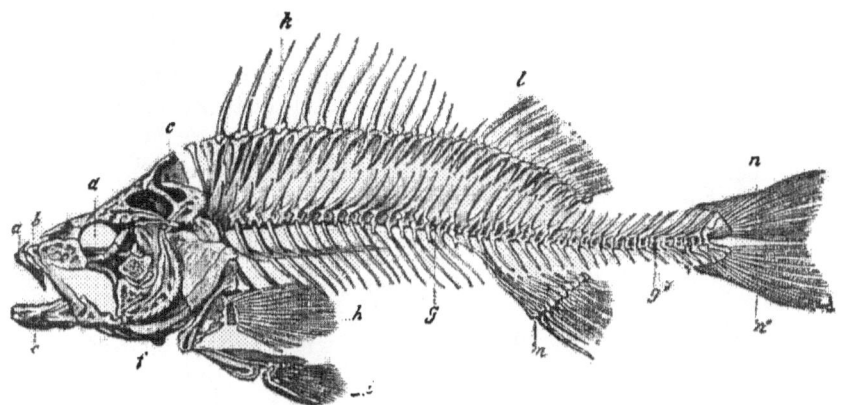

Fig. 49.—Skeleton of the Perch. *a—c*, jaws; *d*, eye; *e f*, portions of skull; *g g'*, backbone; *h*, pectoral fin; *i*, pelvic fin; *k l*, first and second back-fins; *m*, anal fin; *n n'*, tail; *h* and *i* are the paired fins.

the bichir (*Polypterus*) of the upper Nile and the rivers of western equatorial Africa, we should find a totally different structure obtaining in the paired fins. In all these three fishes (of which, be it carefully noted, the first is a representative of the lung-fishes, while the second and third are ganoids) the first pair, or pectoral fins, as is well shown in the bichir, represented in **Fig. 47**, are seen to have a long central lobe running for some distance up the middle of the fin, and completely covered with scales, while the rays of these fins form a kind of fringe, radiating on all sides from the central lobe, the skeleton of a fin of this type being shown in Fig. 50.

From this it will be seen that such a fin consists internally of a long cartilaginous axis, composed of a number of joints (1—9), and that from one or both sides of such joints there are given off obliquely other smaller jointed rods terminating in the fine rays forming the free edges

L

of the fins. How totally different in construction is this kind of fin from that of the perch will be manifest from

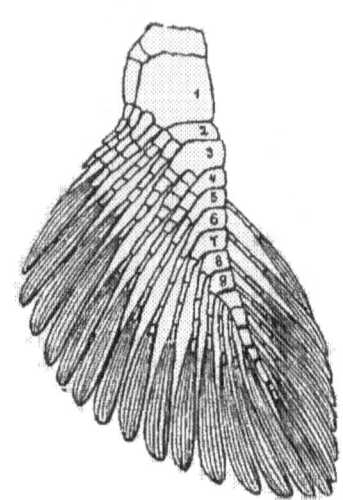

FIG. 50.—Skeleton of Pectoral Fin of an Extinct Shark. (After Fritsch.)

our description and a comparison of Fig. 49 with Fig. 50. Fishes with paired fins like those of the perch may be well termed fan-finned fishes; while those with fins of the type represented in Figs. 47 and 48 may be known as the fringe-finned fishes.

At the present day the only fringe-finned fishes are the Australian lung-fish and the African and American mud-fishes (which, as we have said, are the sole living representatives of the group of lung-fishes), together with the bichir and an allied species, and the gar-pike, which, as we have seen, belong to the ganoids. All the oldest ganoids, such as those found in the old red sandstone of Scotland (Fig. 48), are, however, likewise of the fringe-finned type; and since a gradual passage from these primeval ganoids can be traced through later ganoids found in the upper Palæozoic and Secondary rocks to the bony fishes so characteristic of our present seas and rivers, there can be no manner of doubt but that the fringe-finned type is the most ancient one, and has gradually become modified into the modern fan-finned form.

The evidence that the fringe-finned type is the oldest does not, however, stop here; for, curiously enough, not only had the early ganoids and lung-fishes this kind of fin, but the same type likewise obtained in the primeval sharks. The fin-skeleton represented in Fig. 50 belongs indeed to a member of the same group of sharks as does the species of which the entire skeleton is shown in Fig. 51, where the rod-like axis in both pairs of fins is

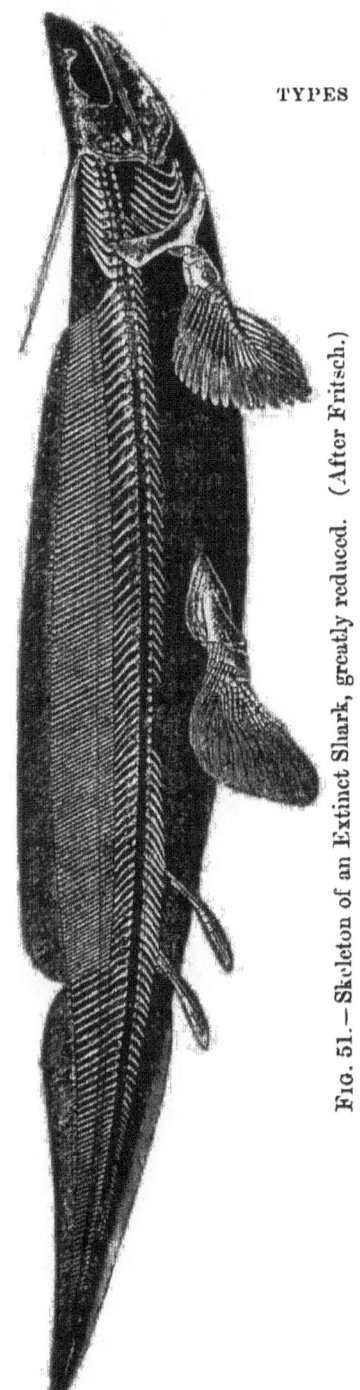

FIG. 51.—Skeleton of an Extinct Shark, greatly reduced. (After Fritsch.)

distinctly seen. Now we have already mentioned that modern sharks and rays have fan-like fins ; and it is, therefore, clear that both in the sharks and in the compound group represented by the ganoid and the bony fishes there has been an independent transition from the fringe-finned to the fan-finned type. On the other hand, in the lung-fishes, which, as we shall see shortly, are a very ancient race, the fringe-finned structure has been preserved without alteration throughout countless ages.

We are still unacquainted with the habits of some of the living fringe-finned fishes, but at least the lung-fishes are species living partially buried in the mud, and are evidently not adapted for swimming rapidly. On the other hand, the fan-finned modern fishes, whether they be sharks or whether they be bony fishes, are generally adapted for rapid motion in the water. Any person who has watched a bowl of gold-fish will not have failed to notice the incessant and rapid motion of their film-like fins, and it is quite evident that this rapid motion could only be produced by fins of the fan-like structure. The fringe-fins

L 2

are, indeed, more like clumsy paddles, capable only of comparatively slow and steady motions; such movements being sufficient for fishes protected either by the bony armour of the ganoids, or by the spines with which the early sharks (Fig. 51) were armed. The fan-like fin is, therefore, obviously the most specialized type of structure, and as the ganoids in their advance towards the bony fishes gradually acquired this fan-like fin, and with it, we may presume, increased speed, it was essential that their enemies the sharks should follow suit in order to be able to catch their prey. This would appear to be a sufficient reason for the attainment of the fan-like structure of fin in both these groups of fishes. It is, however, very remarkable that this structure of the fins having been independently developed in the two groups should have become so alike as it is. On the other hand, the lung-fishes, together with the gar-pike and the bichir, never having had occasion to abandon the mud-loving and sluggish habits of their Palæozoic ancestors, have fortunately preserved for us intact the old fringe-finned type of swimming organs.

It is not, however, in regard to these paired fins alone that fishes show a modification from a long, jointed, axial structure to one which stops suddenly in an expanded termination, from which arises a fan-shaped arrangement of rays, the same kind of modification, although far less generally, having also taken place among fishes in the structure of their tails.

Thus, in all the primeval fishes the backbone (as shown in Fig. 51) is continued right to the very end of the tail, where it terminates in a point. On either side of the backbone are fringes of fin-rays, so that (as shown in Fig 48) in scaled fishes the scaly part of the tail is continued nearly to its extremity. This type of tail is therefore exactly similar in structure to the fringe-finned type of fin, and may be similarly known as the fringe-tailed type. In some fringe-tailed fishes the fringes on either side of the tail (as in Fig. 47) are of nearly equal depth. In other instances, however, the fringe of rays on the lower side is somewhat deeper than that on the upper;

and a further development of this inequality results in the partially-forked tail of the sharks, where the end of the backbone is bent upwards into the upper and longer half of the tail, the lower lobe of which is formed solely of rays. Sharks and lung-fishes have, indeed, never advanced beyond one or other of these two modifications of the fringed-tailed type. On the other hand, the compound group, including the ganoids and the bony fishes, was by no means satisfied with the primitive arrangement of matters. Starting from a fish of the fringe-tailed type like the one represented in Fig. 48, we may trace a gradual shortening of the central part of the tail - fin, accompanied by an increasing development of the rays on its lower side, until we finally reach the completely-forked tail of the perch (Fig.

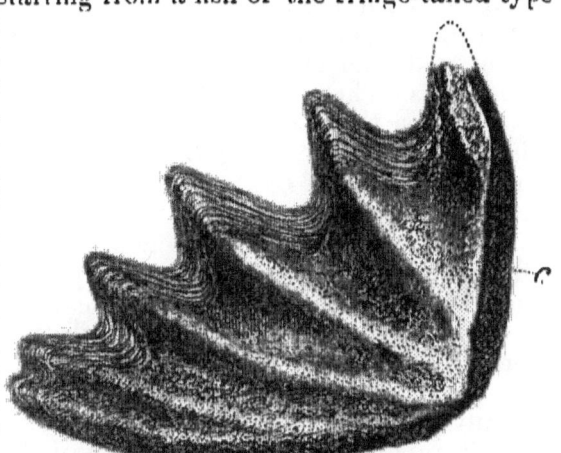

FIG. 52.—Right Upper Tooth of an Extinct Lung-Fish *(Ceratodus).* *C.* point of contact with opposite tooth. (After Teller.)

49), in which, as we have seen, the backbone stops short of the fin-rays, and ends in an expanded extremity from which these rays are given off in a fan-like manner. The bony fishes have, therefore, not only succeeded in developing the fringed-fins of their ancestral ganoids into those of a fan-like type, but have likewise effected a precisely similar modification in the structure of their tails. That the fan-like tail of the perch is an improvement as a steering organ upon the fringed tail of the early ganoids there can be no doubt; and it is such an organ which alone could regulate the movements of the bony fishes in the delicate manner observable in a bowl of gold-fish.   To the sharks, however, whose movements largely partake of vigorous rushes, it is probable that the forked modification of the fringe-tailed type is more advantageous than would have been a tail of the fan type.

Before leaving the subject of the tails of fishes, we cannot forbear to mention that the alteration from the fringed type, with its long central axis formed by the back-bone, from each joint of which springs a pair of rays, to the fan-like type, with all the rays arising together from a blunt and shortened back-bone, is precisely paralleled among birds.   Thus the ancient birds of the Jurassic rocks, known under the name of *Archæopteryx*, had their back-bone prolonged into a long tail, from each joint of which there arose a pair of feathers.   Such a tail was, therefore, essentially a fringed one.   In modern birds, however, as we all know, the backbone extends but a short distance behind the haunch-bone, and then expands into a plough-share-like bone, from which the feathers of the tail radiate in a fan-like manner, very similar to the rays of the tail of a bony fish ; with the exception that whereas in fishes the fan is placed vertically, in birds it is expanded horizontally. In many groups of animals besides these we have mentioned it appears, indeed, that long tails have gone out of fashion, as being useless incumbrances.   We have instances of this in the higher apes and bats, in bears, in guinea-pigs, and in the more specialized kinds of flying-dragons or pterodactyles.

Having said this much as to the fins of the ancient fishes, we may conclude this chapter by giving some particulars relating to the geological history of the Australian lung-fish, which, from the structure of its fins, we have already seen reason to regard as one of the most ancient types of existing fishes.   For a number of years there have been known from the Triassic, or lowest Secondary strata of Europe, fish-teeth of the peculiar type of the one represented in Fig. 52.   The remarkable horn-like form of the ridges on these teeth suggested the name of *Ceratodus* for the otherwise unknown fish to which they pertained.

Nothing more was discovered as to the nature of this problematical fish, and it was even doubtful in what position
these teeth were placed in the mouth, or how many of
them there were in each jaw. Thus matters stood till
some twenty years ago, when naturalists were startled by
hearing that a large fish had been discovered living in the
rivers of Queensland, having teeth like these problematical fossils. This fish was no other than the Australian
lung-fish, which, as we have seen, is one of the few living
species still retaining the ancient fringed fins. It was
found that this fish had one tooth on either side of each
jaw, placed in the same position as the figured example ;
and it was naturally considered that the living fish belonged
to the same genus as the *Ceratodus* of the Trias. Here,
then, we are confronted by the remarkable circumstance
that a kind of fish, first made known to us by fossil teeth
from the very lowest Secondary strata of Europe, was
actually represented by one apparently belonging to the
same genus in the rivers of Australia. It is true, indeed,
that a recent discovery has shown us that the living lung-
fish differs slightly in the structure of its skull from the
fossil *Ceratodus*, and that the teeth of the opposite sides
of the jaws were not in actual contact with one another,
as were those of the latter. Although these points of
difference are considered sufficient to warrant us in regarding the living fish as not actually belonging to the same
genus as the fossil teeth, yet this does not detract from
the extreme interest of the former as being by far the
oldest type of back-boned animal now living. This type
of fish is, indeed, thus proved to have endured throughout
the whole of the immense period during which the entire
series of Secondary and Tertiary rocks were deposited.
And when we reflect that the Secondary rocks include
those enormous accumulations of strata known as the trias,
lias, oolites, greensands, and chalk, while the Tertiary
comprises the threefold divisions termed Eocene, Miocene,
and Pliocene, we can scarcely fail to be almost lost in
wonder at the prodigious length of time during which
lung-fishes have existed, with but comparatively slight
structural modification. The fossil lung-fishes occur in the

Secondary rocks of Europe, Africa, India, and North America, and it is an interesting subject of speculation why the group should have totally disappeared from all those regions, to fin l a last home in far Australia.

The existing Australian lung-fish lives mainly or entirely on leaves, and we may therefore conclude that the fossil teeth likewise pertained to fishes which subsisted on somewhat similar nutriment. If, however, we were to infer that all the teeth of fossil fishes which have a ridged or flattened grinding surface belonged to the herbivorous type, we should be sadly in error, since many of them approximating more or less markedly to the *Ceratodus* type really indicate fishes allied to the well-known Port Jackson shark, in which the mouth is covered with a complete pavement of flattened teeth adapted for crushing shell-fish and other hard animal substances. In all such investigations, the truth can, indeed, only be found out by careful induction, and by availing ourselves of every scrap of information left to us among the relics of former epochs.

CHAPTER XV.

LIVING FOSSILS.

IN the preceding chapter it was pointed out how that, through the discovery of the Australian lung-fish, the ancient Secondary fauna of Europe was brought into much closer connection with that of the present day than had hitherto been supposed to be the case. This, however, is by no means a solitary instance of the discovery in a living condition of forms of life which have been regarded as long extinct; and since the subject of the survival of ancient types in remote corners of the earth or the abysses of the ocean is one of wide interest, we propose to consider it in some detail in the present chapter. For such survivors from a distant past we venture to suggest the title of "living fossils," seeing that for the most part they have but little in common with the dominant fauna of the greater part of the world; while their alliance with extinct types is of the most intimate kind. It is, of course, difficult to know where to draw the line in the use of such an arbitrary designation; but we shall endeavour to restrict the term either to types which, although still more or less abundantly represented at the present day, are of extreme antiquity, or to such as are now represented by comparatively few forms, living either in distant parts of the world or in the ocean depths, but which were abundant in past epochs. Of those coming under the latter category, the majority, as might have been expected, were first made known to science from the evidence of their petrified remains, while their existing relatives were not discovered till later. Whether, however, the extinct or the living types were the first to be discovered, the progress of research has been gradually tending to connect the past more intimately with the present than was originally supposed to have been the case.

Our first examples of "living fossils" will be taken from

the Mollusca, among which the gastropods, described under the name of pleurotomaria, afford the most striking instance. The shells of this genus, which are frequently very elaborately sculptured, have a general external resemblance to a *Turbo* or a *Trochus*, but are readily distinguished

FIG. 53.—Shell of *Pleurotomaria.*

by a deep horizontal slit in the middle of the outer wall of the mouth, from which the genus takes its name. The genus *Pleurotomaria* was originally established in the year 1826 on the evidence of a species from the English lias; and a host of other extinct kinds were subsequently described, ranging from the Silurian to the chalk. Till 1855, no one, however, had the least idea that the genus was still existing, but in that year a living specimen was obtained off Mariegalante, which was subsequently sold in London in 1875 for £25. The species to which this first recent specimen belonged was named *P. quoyana*; and three examples of the same form have subsequently been obtained. The next discovery of a *Pleurotomaria* occurred in 1861, when an imperfect specimen of another species (*P. adansonia*) was obtained. A second example of this same species was taken in 1882 near Guadaloupe, and three others are in existence, while a seventh was purchased in 1890 at Tobago. The latter example was a very large shell, measuring just under six inches in height; and it was also distinguished by its striking coloration, being marked by a number of oblique splashes of reddish-orange on a pale flesh-coloured ground. A third species of the genus (*P. beyrichi*) was obtained from Japanese waters in 1877, and three examples have been subsequently secured. Finally, the fourth and largest living species is only known by a single example, which was recognized in 1879 among a collection of shells at Rotterdam, and is believed to have come from the Moluccas. The height of this fine shell is six and three-quarter inches.

It will thus be apparent that only fourteen specimens of living Pleurotomariæ, referable to four distinct species, are at present known. These molluscs, which inhabit deep water on rocky bottoms, must therefore be extremely rare, although from the nature of their habitat it is probable that not so many specimens are obtained as might otherwise be the case. In the Tertiary period only eleven species are known, of which two are from the Pleistocene, two from the Miocene, and seven from the Eocene. Directly, however, we reach the Cretaceous, the number of species suddenly leaps up to 208, while the total number of Secondary and Palæozoic species is upwards of 1145. Accordingly, out of a total of 1160 representatives of the genus, only fifteen are of post-cretaceous age, of which but four now exist, and these apparently poorly represented in individuals. Here, then, we have indeed a striking instance of a " living fossil."

The well-known bivalve shells named *Trigonia* afford a scarcely less well-marked case of the persistence of early types. This genus was originally named in 1791, on the evidence of an extinct species, but when fully described by Lamarck in 1804 some few recent living examples had also been obtained. For the benefit of those of our readers who may not be familiar with these molluscs, it may be mentioned that the living *Trigoniæ* are rather small shells, of about an inch and a half in diameter, characterized by their somewhat triangular shape, and the strongly-marked transverse ribs, marked with rough tubercles on the outer surface. Internally the shell has a polychroic pearly lustre ; while the peculiarly-shaped and striated interlocking hinge, when once seen, can never be mistaken. At the present day the *Trigoniæ* are represented only by some five closely allied species or varieties, confined to the Australian seas ; while in the Tertiary, although more widely distributed, they were likewise rare. In the Secondary period, where they range down to the lias, these shells were, however, extremely abundant, and attained far larger dimensions than their existing relatives. Indeed, in the oolites *Trigoniæ* were some of the most common molluscs, whole slabs of rock being sometimes found paved with their handsomely sculptured shells, while all who have visited the Isle of Portland must be familiar

with the countless swarms in which casts of their shells
occur in the so-called "roach-bed" of the quarrymen.
The survival of *Trigonia* in the Australian seas alone
affords a curious parallel to the persistence of pouched
mammals and monotremes on that island-continent, and
of the lung-fish in its rivers.

The pearly nautilus, of which there are some three or
four species from the warmer seas, is likewise entitled to
occupy a place among "living fossils," since this group of
cephalopods has existed continuously to the present day,
from the epoch of the lower Silurian, with a progressive
diminution in the number of species. It was long considered that the Palæozoic nautili were congeneric with
the existing ones, but although this is probably not the
case, the whole are so closely allied as to show a most
remarkable persistence of type throughout untold ages;
nautili have, indeed, witnessed the incoming and the
decline of the whole group of ammonites, so characteristic
of the Secondary rocks; but the reason of the persistence
of the one type and the total extinction of the other type
appears entirely beyond our ken.

The most remarkable instance of the persistence of type
is, however, afforded by the genus *Lingula* among the
brachiopods, or so-called lamp-shells. Lingulas have
oblong, flattened, and somewhat nail-like shells, composed
partly of horny and partly of calcareous matter, and are
attached to foreign substances by a long muscular pedicle
passing out between the beaks of the two valves, which
are generally of a greenish hue. These molluscs range
from the Cambrian—at the very base of the Palæozoics—
to the present day, without any trace of generic modification, and indeed with no perceptible change. Moreover,
the group seems to be now as well represented in species
as ever it was, the total number of living forms being
given in the second edition of Woodward's "Mollusca"
as sixteen, while the total of fossil species at that date was
but ninety-one. The lingulas are, therefore, the very
oldest animals at present in existence. They are, however, run somewhat close by two other genera of brachiopods respectively known as *Discina* and *Crania*, both of

which range from the lower Silurian or Ordovician epoch
to the present day. The parallelism in this respect is not,
however, so close as it might at first sight appear, since the
lower Palæozoic representatives of both the latter genera
are subgenerically distinct from their living analogues.
While the first of these two genera has some ten living
species, the latter possesses but five, and both had a large
number of Palæozoic representatives.

It is, perhaps, almost superfluous to add that the whole
of the brachiopods are a waning group, although the types
known as rhynchonellas and terebratulas have been ascertained, of recent years, to be more numerously represented,
both as regards species and individuals, than was formerly
considered to be the case.

The so-called stone-lilies, or crinoids—near relatives of
the familiar star-fish, but attached, in the young condition
at least, to the sea-bottom by a jointed stem—likewise
constitute a group in which by far the greater number of
types are totally extinct, although a few survive to merit
the title heading the present chapter. The stone-lilies are
divided into two primary groups, one of which is totally
extinct, while the other, which does not extend backwards
beyond the Secondary period, comprises the few existing
representatives of the class. The genus which may be
selected as especially worthy of the designation "living
fossil" is *Pentacrinus*, so named from the pentagonal form
of the discs of the stalk—so familiar to all who have
studied the fossils of the British Secondary rocks. The
pentacrinids were originally named on the evidence of
certain species from the British lias, which attained a
height of several feet, and flourished in extraordinary
profusion on the old sea-bottom, as the reader who cares
to pay a visit to the fossil galleries of the Natural History
Museum may see for himself. For some time the group
was only known from the Secondary rocks, but eventually
a minute species (*P. europæus*), supposed to belong to it,
was found living in deep water off the Irish coast; while
in 1755 a much larger form (*P. caput-medusæ*) was discovered in the West Indian seas. The former turned out,
however, to be merely the larva of the familiar feather-star

(*Antedon*) ; although the latter, together with many other subsequently discovered species, was a true crinoid. For a long period these living crinoids were supposed to be of

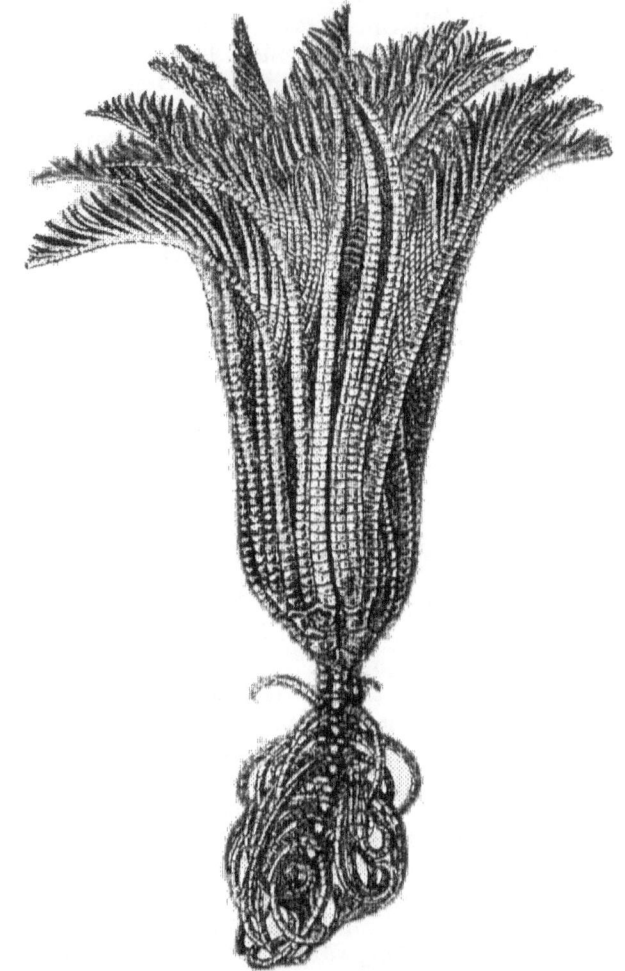

Fig. 54.—A Living Pentacrinid.

extreme rarity, only a few examples being from time to time procured. With the development of deep-sea dredging it was, however, found that pentacrinids were really abundant in certain localities, and several new species (one of which is represented in the accompanying figure) have been named of late years. Thus in the summer of 1870, Gwyn Jeffreys dredged up quantities in the Atlantic; while in the neighbourhood of the Aru Islands Moseley tells us that during the voyage of the *Challenger* more than thirty specimens of Pentacrinus were taken at a single haul of the dredge in 500 fathoms of water. Although their pyritized remains, which are so common on the slabs of lias, would have indicated that these crinoids must have been creatures of extreme beauty, no. adequate idea of their gracefulness would ever have been obtained had the group not been represented in the living state.

Pentacrinids are, however, by no means the only "living fossils" belonging to this class of animals. For instance, the smaller and simpler *Rhizocrinus* of the Atlantic is a survivor from the Eocene; while both this genus and the allied *Bathycrinus*, from depths reaching to 2400 fathoms in the Atlantic, are near relatives of the extinct *Bourgueticrinus* of the chalk. Moreover, all these forms are related to the so-called pear-encrinites (*Apiocrinus*), so common in the middle Jurassic rocks of Europe; while these, again, lead on to the still earlier lily-encrinites (*Encrinus*) of the trias.

If space permitted, there are several other groups of invertebrates which might claim our attention, but we must pass on to the vertebrates. Among the fishes the one which has the greatest claim to the title of a "living fossil" is the aforesaid Australian lung-fish of the rivers of Queensland. As mentioned in the last chapter, teeth of fishes allied to the Australian lung-fish have been long known from the Secondary rocks, ranging downwards to the trias, and occurring in Europe, India, Africa, and North America. These teeth were, indeed, first described by Agassiz as far back as the year 1838; and the group was believed to be extinct till 1870, when one of the two living forms was discovered. At first, as stated in the same chapter, it was believed that the latter were generically identical with the fossil. *Ceratodus,* but it has been ascertained recently that there are certain slight differences which justify the separation of the living species as a distinct genus. At present we have no decisive evidence of the existence of these fishes between the upper Jurassic of the United States and the Pleistocene of Queensland, so that there is a long gap in their history to be filled up, it may be hoped, by future discoveries.

The Australian lung-fish is, however, but one of three nearly allied genera, of which the other two (*Lepidosiren* and *Protopterus*) are each represented by a single species, severally inhabiting the Amazon and the rivers of West Africa. These three widely separated forms are, then, the sole living representatives of an extensive order which was once widely distributed over the globe, and has been slowly waning ever since the Palæozoic period. The group is one of especial interest, since from some of its extinct representatives, or nearly allied fishes, it is probable that amphibians, and hence the higher vertebrates, have all been derived.

Of nearly equal interest with the lung-fishes are the bony pikes (*Lepidosteus*) of the rivers of North America and the bichir (*Polypterus*)* of the upper Nile and the rivers of Western Africa; which, together with another West African form (*Calamoichthys*), are the sole existing representatives of the mail-clad ganoid fishes so abundant during the Secondary period. The African forms are at present unknown in the fossil state, but the bony pikes date from the lower Eocene, and thus indicate continuity with the extinct Secondary types.

Another living fossil among fishes is the well-known Port Jackson shark (*Cestracion*), the last survivor of a genus ranging in the Secondary rocks of Europe down to the Kimeridge clay; and also the sole living member of a vast group of sharks characterized by the pavement of crushing teeth with which the mouth is covered. As remarked long since by Buckland, cestraciont sharks lived side by side with trigonias in the old Jurassic seas of Europe; and it is not a little remarkable to find the

---

* See page 145.

same comradeship still kept up on the distant coasts of Australia.

Quite recently another link connecting the present fauna of Australia with that of Secondary Europe has been discovered. For a considerable time a peculiar group of herrings (*Diplomystus*), characterized by having a row of scutes on the back resembling those found in other types on the opposite aspect of the body, have been known from Cretaceous and early Tertiary rocks, their range including Brazil, Wyoming, the Isle of Wight, and the Lebanon. Till the other day, these doubly-armoured herrings were considered to be totally extinct, but now, lo and behold! they have turned up alive in certain rivers of New South Wales.

Among amphibians, the creature which seems best entitled to be called a "living fossil" is the giant salamander (*Cryptobranchus*) of Japan, since, together with a smaller North American kind, it is the representative of a genus once common in Europe during the middle portion of the Tertiary period. Indeed, our first knowledge of the group was derived from a fossil specimen of one of these salamanders from the continent, described in the year 1726 under the title of *homo diluvii testis*, in the belief that it was a human skeleton!

Passing on to the reptilian class, we have to notice that in the year 1842 Sir R. Owen described from the Triassic rocks of Shropshire the remains of a small lizard-like reptile (*Rhynchosaurus*), differing from all living forms in the structure of its skull, of which the jaws terminated in a peculiar beak. Eleven years previously Dr. Gray had, however, applied the name *Sphenodon* to a then very imperfectly known living reptile from New Zealand; which when fully described by Dr. Günther in 1867 turned out to be very closely related to the Triassic *Rhynchosaurus*. Although externally somewhat like a lizard, but with a different kind of skin, the tuatera, as the New Zealand reptile is called, differs entirely in the structure of its skull and skeleton in general from the true lizards, and comes much closer in these respects to crocodiles and tortoises. Subsequent researches have

M

brought to light the existence of the remains of a large
number of more or less nearly allied reptiles in the
Secondary rocks of Europe and other parts of the world.
Accordingly the tuatera, which, although not generically
identical with any one of these extinct forms, has every
right to be regarded as a " living fossil "; while it enjoys
the further distinction of being, with the exception of
the lancelet, the only vertebrate animal which can be
definitely regarded as the sole living representative of a
distinct order.

Since birds have no species with any very great claim to
be mentioned here, we pass on to mammals, of which our
notice must necessarily be brief. In a previous chapter
it has been stated that certain remarkable Secondary
European and American mammals appear to be related
to the egg-laying mammals of Australia and New Guinea ;
and we may, therefore, assume that the latter, as their
structure indicates, are very ancient types, although their
direct ancestors have not yet been discovered. The banded
anteater (*Myrmecobius*) and the bandicoots (*Perameles*) of
Australia seem to be the nearest relatives of another great
group of Secondary mammals, and are therefore probably
some of the oldest types with which we are yet acquainted,
although here again their exact genealogy is at present
unknown. No other groups of living mammals are yet
definitely known to have existed before the Tertiary period,
and the pedigree of the class in general is consequently
brief as compared with that of many of the animals
discussed above. The opossums (*Didelphys*) are, however,
perhaps those mammals best entitled among the Tertiary
groups to the appellation of "living fossils," as they have
existed without generic modification since the period of
the Eocene, and have now entirely vanished from their
old European haunts to maintain an existence in America,
where they are mainly characteristic of the southern half
of the continent. Although the insectivores and the
lemurs are evidently primitive types, but few of their
existing genera date far back in the Tertiary period, while
in the latter group not a single existing genus is known
before the present epoch. None of these mammals

properly come, therefore, within the scope of the present chapter. On the other hand, tapirs and rhinoceroses, as dating from the lower part of the Miocene or the upper portion of the Eocene period, might be considered to claim notice in our survey, while the same remark would apply to the civets of the genus *Viverra*. Since, however, these mammalian types are comparatively well represented at the present day, they scarcely come under the designation of "living fossils." There is, however, one mammal to which this title is strictly applicable, namely, the water-chevrotain of Western Africa. This genus of mammals was originally made known to science upon the evidence of fossil remains from the Pliocene rocks of Darmstadt described under the name of *Dorcatherium* in 1836. Four years later, a living ungulate from West Africa was described as a species of musk-deer (*Moschus*), and the same creature was in 1845 made the type of a new genus, *Hyomoschus*. Subsequently other extinct species of *Dorcatherium* were described from the Miocene rocks of Europe and the Pliocene of India, and it was eventually proved that the African water-chevrotain belonged to the same genus as these reputedly extinct forms. Hence, the animal in question, as being the sole existing representative of a genus which had formerly a comparatively wide distribution and which was originally described as extinct, has the most indisputable claim to rank as a "living fossil."

CHAPTER XVI.

THE EXTINCTION OF ANIMALS.

IF it be true that, as compared with the pre-glacial epoch, we are living in an impoverished world so far as the larger forms of animal life are concerned, it is even more certain that our immediate descendants will be the heirs of a still more depopulated globe. Already the bison has disappeared from the American prairies, while of the vast swarms of antelopes and other large mammals which half a century ago peopled the plains of Africa scarce a tithe remains, and a few are well-nigh, or even totally, extinct. A few years ago, indeed, it appeared that we were likely ere long to witness the complete extermination of many species of the larger African mammals; but, although we can never expect to see them again in their original multitudes, several of the South and East African governments are now taking measures to ensure the preservation of a few of the various species, and there is accordingly some hope that although the nineteenth century will ever deserve the reproach of posterity, as having been the one during which the world was to a great extent depopulated of the larger forms of animal life, yet that it will escape the crowning shame of having actually exterminated a host of species. Still, however, the indictment against our generation is heavy enough in all conscience; and there is no reasonable doubt but that the destructiveness which is so characteristic of human nature—whether civilized or otherwise—has led to the total extinction of two species of large African mammals within the last thirty years, while a third is only too likely to share the same fate. On the other hand, it must not be assumed that all the animals known to have become extinct within the historic period have succumbed directly to this demon of destructiveness. In certain cases, as we shall see in the sequel, the introduction of other animals by human agency has been the involuntary means of leading to the final extinction ; while occasionally, as in the case of the great auk, a catastrophe of nature has accelerated the climax. In other instances, it would seem that a species has been gradually dying out from unknown natural causes, in the manner which appears to be natural to all forms of animal life, where species and genera, like individuals, have but a certain allotted span of existence.

In the present chapter we propose to notice the chief animals, exclusive of invertebrates, which are known to have been exterminated, or are just verging on extinction, within the historic period ; but before doing so we may briefly allude to a few others which urgently require protection, unless they are to share the same fate. Foremost among these are the African elephant and the so-called white, or square-mouthed, rhinoceros of the same country. Till within a short time ago there were, indeed, strong grounds for believing that the latter magnificent animal, which at present is represented only by a few skulls and horns in English collections, had already disappeared ; but we are rejoiced to hear that a few individuals still linger on in a remote corner of Eastern Africa, where it is to be hoped they will receive immediate and adequate protection. The walrus of the polar regions is another animal whose numbers have been woefully diminished of late years, and which likewise stands in urgent need of protective legislation, while a similar remark will apply to several species, or local races, of seals. In countries like New Zealand, where there were originally no native carnivorous animals, many of the indigenous creatures now stand in great jeopardy by the introduction of the latter, and it is only too likely that the flightless kiwis of those islands will eventually be exterminated by half-wild cats and dogs. The curious tuatera lizard of the same country is also likely to be killed off by pigs.* In the Samoan Islands a similar fate long threatened the tooth-billed pigeon (*Didunculus*)—the nearest living ally of the dodo—but the impending destruction was fortunately averted by the

---

* See page 161.

bird having forsaken its original habits and taken to
perching on trees.

Strictly speaking, the moas of New Zealand come within
the category of animals exterminated within the historic
period, since they were almost certainly killed off by the
Maories ; but as we have no direct historic evidence of
their existence, no further mention of them will here be
made. We shall commence our survey with three species
which were the first to succumb.

When the Dutch Admiral Van Neck visited Mauritius
in the year 1598, he found that island inhabited by a
number of ungainly, flightless birds, which he called
walghvogel (disgusting fowl), but which were afterwards
termed by the Portuguese dodo (from *doudo*, a simpleton).
Subsequently, many living examples of this bird (the form
of which is probably well known to our readers) were
exhibited in Holland, and their portraits painted by the
two artists Savery. The museum at Oxford also once
possessed a stuffed specimen, which, with the exception of
the head and a foot, were in 1755 destroyed, as being
too much decayed to be worth preserving! In the year
1601 a Dutch ship captured twenty-four dodos for pro-
visions, and others soon followed suit, so that at this time
the bird was still common. It had, however, disappeared
before the close of the century. The last notice of the
living bird occurs in the journal of the mate of the *Berkley
Castle* in 1681 ; and from the absence of any mention of
it by Leguat, who visited the island in 1693, we may
presume that it was already quite extinct. Although the
numbers carried away by ships doubtless largely aided in
its extermination, Prof. Newton is of opinion that the
dodo was finally killed off by the pigs which had run
loose over the island. There is evidence that an allied,
although totally unknown bird formerly inhabited the
neighbouring island of Réunion.

Near akin to the dodo was the taller and more lightly-
built pigeon-like bird formerly inhabiting the island of
Rodriguez, and known as the solitaire (*Pezophaps*). Our
sole knowledge of this bird in a living state is derived
from the accounts of Leguat, who founded a colony on the

island in 1691, and who has left us not only a good account of its habits, but likewise an excellent portrait. When the solitaire became extinct is uncertain, but there is some evidence that it may have lingered on in the more remote parts of the island down to the year 1761. These birds were flightless, and the males much larger than the females, their rudimentary wing having a peculiar horny ball-like excrescence. Up to the year 1864 our museums possessed only a few bones of these strange birds, obtained from caverns in Rodriguez ; but during the transit of Venus expedition a large number of remains were obtained, from which several more or less nearly perfect skeletons were set up.

Mauritius and Rodriguez also possessed another large flightless bird, known as the *Aphanapteryx*, and belonging to the rail family. By the fortunate discovery of an old painting, we learn that this bird had a long recurved bill and a brownish-red plumage. It was living in 1615, but seems to have disappeared by Leguat's time.

Another extinct Mauritian bird was the géant (*Leguatia*), which was a kind of coot, described by Leguat in 1695 as being equal in size to a goose. When it died out is unknown. A remarkable crested parrot (*Lophopsittacus*) —the sole representative of its genus—also existed in Mauritius in 1601, which has long since disappeared.

Our next instance of extermination relates to a very different kind of creature, viz., the great northern sea-cow (*Rhytina*), a near ally of the existing dugongs and manatis of the warmer seas. The rhytina, which was far larger than its living cousins, attaining a length of from twenty-four to thirty feet, was discovered by the ill-fated navigator Behring, on the island which bears his name, in the year 1741 ; and had it not been that he was accompanied by an excellent naturalist (Steller), it is quite probable that the creature might have perished without our ever having even heard of its existence. This sea-cow was confined to Behring and Copper Islands at the date of its discovery, where it existed in large numbers ; but there is little doubt that it must formerly have had a much wider range, and that it was even then a waning race. Although

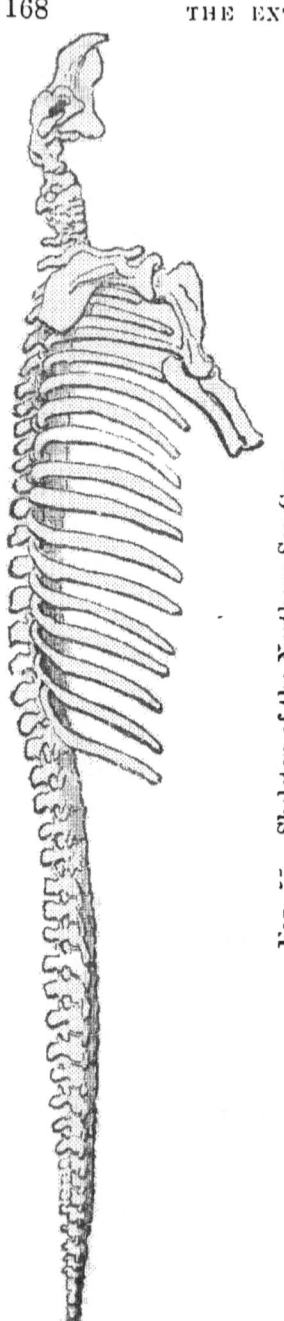

FIG. 55.—Skeleton of the Northern Sea-Cow.

Behring and his party were wrecked on Behring Island, where they remained for some ten months, it does not appear that they inflicted much damage on the sea-cows; but soon after, various fur-hunting expeditions were fitted out to Alaska, the members of which depended solely on these animals for food. So incessant, indeed, was the persecution of the unfortunate creatures, that by 1754 they had been exterminated on Copper Island, while by 1763 nearly all had been killed on their other haunt, and the last individual of their race is supposed to have perished either in 1767 or the following year. Up to 1883 three skeletons in foreign collections and a few ribs in the British Museum were all that remained of the rhytina; but since that date numerous remains have been obtained from the peat of Behring Island. Fortunately, the excellent description left by Steller, and some drawings which have come down to us from the navigator of Behring's party, give us a very good idea of the external form of the giant sea-cow.

Continuing our chronological survey, the next exterminated animals that claim our attention are the gigantic land tortoises of the Mascarene Islands, that is to say Mauritius, Rodriguez, and Réunion. In Mauritius these tortoises were first discovered by Van Neck, the discoverer of the dodo, in 1529, who relates that some of them were of such huge dimensions that six men could be seated on their shells. In Réunion, the voyager Boutehoe writes that he took twenty-four giant tortoises from beneath a single tree, in the year 1618; while Leguat, in 1691, states that in Rodriguez "there are such plenty of land-turtles in this isle that sometimes you see two or three thousand of them in a drove, so that you may go about a hundred paces on their backs." The Réunion tortoises, of which not a single specimen remains in our museums, seem to have been the first to disappear, although at what date is uncertain. A solitary individual apparently, however, survives at Mauritius, where it has lived for considerably more than a century, having been imported from the Seychelles, where it was doubtless carried from Réunion. Down to the year 1740 these reptiles were still abundant in Mauritius, but by 1761, when vessels were employed in transporting them to that island from Rodriguez as food, they had probably become scarce; while in both islands the whole race became extinct early in the present century, mainly owing to the ship-loads which were carried away for food; although rats and pigs have largely aided in the work of destruction by devouring the eggs and young. It may be added that giant tortoises were formerly widely distributed over the world; but that within the historic period they have existed only in the Mascarenes, in Aldabra, to the north of Madagascar, and in the far distant Galapagos group. In Aldabra, whence they have been introduced into the Seychelles, they are now becoming very scarce, although we believe they receive some kind of protection. The Galapagos tortoises are, however, only too likely to share the fate of those of the Mascarenes, most of the larger specimens having been already killed off. Indeed, it is highly probable that the Abingdon tortoise, remarkable for the extreme thinness of its shell, specimens of which were obtained in 1875 for our national collection, may already be numbered with the extinct.

With the pied starling (*Fregilopus varius*) of Réunion we return once more to birds. This beautiful species, the sole representative of its genus, and distinguished not only by its pied plumage but likewise by the presence of a crest of feathers on the head, is said to have been very bold and confiding in its disposition, and is believed to have been exterminated about fifty years ago.

Within the last few years a species of ke-ke parrot (*Nestor productus*), formerly inhabiting Phillip Island, near New Zealand, is likewise believed to have become extinct about the middle of this century.

Our next species is one which may possibly be still existing, although, if so, it must be of extreme rarity. This is the gigantic blue coot (*Notornis mantelli*) of New Zealand, which is almost the sole representative of its genus, although it has near allies in the purple water-hens (*Porphyrio*). This huge flightless bird was first made known to science on the evidence of some bones obtained from the volcanic sands of Waingongoro, in the North Island, and described by Sir R. Owen as belonging to an extinct form. Their discoverer, Mr. W. Mantell, succeeded, however, in obtaining the skin of an example which had been caught alive and eaten by some sealers in the South Island, some time in 1877; this skin, together with that of the second specimen, being now mounted in the British Museum. The second living specimen was taken in 1869, and a third in 1881, both in the South Island; but whether any others still survive is more than doubtful. As the fossilized remains of this species are not uncommon in the superficial deposits of both islands, we may probably refer its extinction in the North Island, and its extreme rarity in the South, to the Maories. If it still linger it is probable that wild pigs, dogs, or cats will ere long put a term to its existence. An allied species (*N. albus*), distinguished by its plumage, formerly inhabited Norfolk and Lord Howe Islands, but is now extinct.

The great auk, or gare fowl, brings us to a species completely exterminated in modern times, and of which the accounts are fairly complete. This bird, the largest member of its genus, and totally unable to fly, was restricted to the shores of the North Atlantic, ranging in Europe from Iceland in the north to the Bay of Biscay in the south, while in America it extended from Greenland to Virginia. These southern limits mark, however, only the winter range of the species, which was somewhat migratory in its habits. Its breeding-places were but few, the chief being the rock called Geirfuglasker off the coast of Iceland, and Funk Island on the Newfoundland coast; both these spots being bare, barren rocks very difficult of access. In spite of its slow increase (but a single egg being laid at a time), the great auk existed in countless numbers on Funk Island, where it was discovered by Cartier in 1534. Here for nearly two centuries it formed an unfailing food-supply for all vessels visiting the neighbouring seas; and it might have lived till now had not the custom arisen of men being landed on the island to spend the summer in slaying these birds for the sake of their feathers. It is said, indeed, that the auks were actually killed by millions, being first driven into stone enclosures, and then bludgeoned. When the bird disappeared from the American side is not quite clear, although it was probably somewhere about the year 1840. Four years later it had also ceased to exist on the opposite side of the Atlantic, the last European pair having been killed in the summer of 1844. What led to its rapid and final extermination in Europe was the sudden subsidence of the Geirfuglasker in 1830, which compelled the birds to seek other and more accessible breeding-places, where they were less protected from molestation. The last British example was killed in Waterford harbour in 1834. In addition to bones obtained from Funk Island, the great auk is now represented in collections only by some seventy-six skins, nine skeletons, and sixty-eight egg-shells.

Before noticing the few remaining birds on our list, we may refer to two South African mammals which are now almost certainly extinct. The first of these is the fine antelope known as the blaubok (*Hippotragus leucophæus*), a near ally of the handsome sable antelope and roan antelope, characterized by their large scimitar-like and backwardly-sweeping horns. Indeed, the roan antelope is often confounded with the blaubok, although the latter was a considerably smaller and otherwise different species. When the blaubok was killed off cannot now be ascertained, but it was certainly abundant at the Cape in the first half of this century. Unfortunately, the British Museum has not a single specimen of this antelope, although a head is preserved in Paris.

The second African mammal is the quagga (*Equus quagga*), a near relative of the zebras, but distinguished by the hinder portion of the body being devoid of stripes. This animal was described by Sir Cornwallis Harris in 1839 as existing in immense herds, although its distribution was always very local; but of late years there is no definite record of a single specimen having been seen. If, as is probably the case, it is truly extinct, there is no record of the date of its disappearance. Of the almost total extermination of the square-mouthed rhinoceros, mention has been already made.

Curiously enough, the northern sea-cow was not the only animal discovered on Behring Island in 1741, during Behring's involuntary sojourn, which appears to have since become extinct. This second species was Pallas's cormorant (*Phalacrocorax perspicillatus*), the largest representative of its genus, and distinguished by its lustrous green and purple plumage, and the bare white spectacle-like rings round the eyes. This bird, which weighed from twelve to fourteen pounds, had small wings and was a poor flyer, with a stupid, sluggish disposition. Steller relates that it occurred in great numbers, and was extensively used as food by the members of Behring's party. About 1839, Captain Belcher, of the "Sulphur," received as a great rarity a present of one of these fine birds from the Governor of Sitka, by whom some other specimens were sent to St. Petersburg; but since that date nothing has been heard of the species, which probably became extinct within about a century of its discovery. Dr. Stejneger, who visited Behring Island in 1882, instituted a careful search after this bird in vain, although he was rewarded by finding some of its bones buried in the soil. The species is now represented only by four mounted specimens, one of which is in our own national collection, and a few bones.

Another bird that appears to have become extinct within the last half-century is the beautiful black and golden sickle-bill, or mamo (*Drepanornis pacifica*), first brought to Europe by Captain Cook after his discovery of the Sandwich Islands, to which group it was restricted. This bird belonged to the family of honey-suckers, and was remarkable for the length of its curved bill. The brilliant yellow feathers from the back of the bird were used by the Hawaiian chieftains in the manufacture of their gorgeous feather-cloaks; and as one particular cloak, according to Mr. Scott Wilson, measures four feet in length and more than eleven feet round the base, it may be imagined what a number of birds of eight inches in length—and these only yielding the particular feathers on the back—would be required for its construction. Indeed, the manufacture of this particular cloak is stated to have lasted through the reigns of eight chieftains; and it is to the destruction thus caused that Mr. Wilson attributes the extinction of this beautiful bird, now represented in our museums only by some four stuffed examples.

In the Antilles a remarkable burrowing petrel, known as the diablotin, is also believed to be now extinct.

With the handsomely marked Labrador duck (*Somateria labradorias*), a near ally of the eider-duck, we bring to a close our list of animals which can be pretty definitely affirmed to be extinct, although there are a few others over which the same fate is impending, even if it has not already befallen them. This duck was not unlike the common long-tailed duck (*Harelda glacialis*) in general coloration and size, although without the long tail of the latter. In the male, the body and primaries are black, and there is also a ring round the neck and a stripe down the head of the same hue; while all the remainder is white. During the breeding-season this species inhabited Labrador, but in winter its range extended as far south as New Jersey. According to Mr. F. A. Lucas, to whom we are indebted for much information concerning the extinction of several of the species noticed in this chapter, the

Labrador duck was never common, and as no example has been seen since 1879, it may fairly be presumed to be extinct.

In conclusion, we may refer to a very remarkable mammal of which but a single example has hitherto come under civilized human ken. As most of our readers are probably acquainted, at least by name, with the large spotted South American rodent, known as the paca—a distant cousin of the familiar guinea pig—we shall assume that they know what we mean when we talk of a paca-like animal. Now, on a certain occasion, somewhat more than twenty years ago, the inhabitants of Montana de Vitoc, in Peru, were surprised to find at daybreak a large rodent with the general appearance and coloration of a paca walking unconcernedly about the courtyard of a house. The creature differed, however, from a paca in having a tail of considerable length, as well as in its smaller ears and cleft upper lip; while dissection revealed other internal points of distinction. The curious part of the matter is that none of the natives of Peru had ever previously heard of or seen a similar animal (for which, by the way, the name of *Dinomys* was suggested), and from that day to this nothing more has been heard about the creature. Can it be that the specimen then seen and killed was the last survivor of its race, and that the *Dinomys*, whose existence was thus strangely revealed to us, must also be numbered with the extinct?

CHAPTER XVII.

PROTECTIVE RESEMBLANCE IN ANIMALS.

THAT the colours of animals tend to assimilate themselves to the natural surroundings of the animals themselves is a fact which has long been known in natural history; and it is, indeed, one which is self-apparent to every sportsman and to every traveller in the wilder regions of the globe. For instance, everyone is probably aware that desert-haunting animals, like lions, gazelles, wild asses, jerboas, and many species of birds, generally have a uniform sandy-coloured coat, which renders them at a short distance almost or completely invisible in their native wastes. Then, again, every English sportsman knows how completely the coloration of the partridge and the hare assimilates with that of the stubble or ploughed fields in which they are wont to lie; while the mottled blacks and browns of the woodcock and snipe accord so exactly with the hues of decaying leaves and grass that the inexperienced eye will often fail to detect a wounded bird even when lying close to the feet, and scarcely anyone can distinguish the living birds when on the ground. The brilliant vertical orange and black stripes of the tiger and zebra, when seen in a menagerie or a museum, do not strike us as resembling anything in inanimate nature. In its native jungle, largely composed of upright yellow stems of tall grasses, between which are narrow intervals of deep black shade, the colour of the tiger is, however, admirably suited to its surroundings; and it is stated that the stripes of the zebra are arranged in such proportions as exactly to match the pale hue of arid ground by moonlight, so that on such occasions these animals are absolutely invisible even at very short distances, while they are by no means easy to distinguish at a moderate distance even in broad daylight. We hardly need refer to the white colour of polar animals, such as bears, ermines, foxes, hares, &c.,

as the most perfect example of this kind of protective
coloration; and numerous other examples will at once
present themselves to the reader.

Well known as are these comparatively simple instances
of protective resemblances, there are, however, others of
a more striking nature, where the animal either resembles
the form of some inanimate object, or that of some other
kind of animal which has especial means of protection;
and since these resemblances are less generally known,
they will form the subject of the present chapter. The
term "protective resemblance" is generally applied to those
instances where the animal resembles more or less closely
an inanimate object, and thus renders itself inconspicuous;
while the instances where one animal assumes the appearance of another, and thereby becomes conspicuous, are
classed under the term "mimicry." It will, however, be
obvious that both these kinds of resemblances are near
akin, and are far in advance of protective coloration, pure
and simple, where no imitation of form takes place. We
shall first mention some instances of the imitation of the
forms of inanimate objects by animals, and then refer
to those cases where other animals are the objects of
imitation.

Some of the best examples of what we shall take leave
to call inanimate mimicry are to be found among insects,
and we shall take our first case from among the butterflies. All are probably aware that a large number of these
insects, such as our common peacock and tortoiseshell
butterflies, while brilliantly coloured on the upper surfaces
of their wings, have the under surfaces of the same of a
dull, sombre hue; and most of us have doubtless been
almost startled at the suddenness with which one of these
gaudy creatures seems to vanish altogether when it settles
on dark ground or the rough bark of a tree, and at once
closes its wings. Here, then, we have an instance of
ordinary protective coloration, without any attempt at
mimicry of form. There is, however, a peculiar group of
butterflies allied to our own purple emperor, inhabiting
northern India and the Malayan region, which have gone
far beyond this simple kind of protective resemblance,

and actually imitate the form of leaves growing on their native branches. These butterflies (scientifically known as *Callima*) have the upper surface of the wings brilliantly marked with orange, their front wings terminating in a sharp point externally, and the hind ones in a "tail," after the fashion of our swallow-tailed butterflies. Between the sharp point of the front and the tail of the hind wing there runs on the under surface a curved line, from which smaller lines are given off to the edges of the wings. When this butterfly settles on the stem of a plant bearing pointed leaves, and closes its wings, the points and tails of the same of course come into exact opposition ; and since the tail of the wings is closely applied to the stem of the plant it appears exactly as though it were the stalk of a leaf, the midrib and veins of which are exactly imitated by the lines on the under surface of the wings ; while the apex of the leaf is formed by the opposed points of the front wings. So exact is the resemblance of the butterfly when in this position to a faded leaf, that, as Mr. Wallace tells us, it deceives the eye even when gazing full upon it, and without actually seeing the insect settle upon the spot it is absolutely impossible to find it. To increase the delusion no two individuals of these insects are precisely alike on the under surface ; while many of them have little black patches, or dots, exactly resembling the dark fungous growths so often found on decaying leaves. Fortunately for the reader who desires to verify this extraordinary instance of inanimate mimicry, a case is now exhibited in the central hall of the Natural History Museum with several of these insects attached to a bough with faded leaves, and it is curious to watch the visitors to this case and see how often they fail to distinguish all the butter-flies from the leaves which they imitate.

The next best instance of inanimate mimicry among insects occurs in the so-called stick- and leaf-insects, which are allied to our grasshoppers and cockroaches. The stick-insects, some of which are found in southern Europe, have long, slender bodies and limbs, of a dark colour, and so exactly resemble dry sticks that it is almost impossible to distinguish between the one and the other.

N

To increase the resemblance, these insects when at rest have the habit of placing their legs unsymmetrically. On

FIG. 56.—The Leaf-Butterfly. (After Wallace.)*

* The author is indebted to Messrs. Macmillan & Co. for this figure.

the other hand, the leaf-insects, or "walking leaves," of India, both in colour and form, so exactly simulate green leaves that they may be passed dozens of times without attracting attention. All the legs of these curious creatures are furnished with irregular flat expansions looking precisely like bitten leaves ; while the head and fore part of the body forms a kind of stalk expanding behind into a broad and flattened abdomen, covered by the horny wings, which are veined and netted so as to form an almost exact imitation of a leaf.

We might cite many other instances of inanimate mimicry among insects, but we must pass on to show that this phenomenon is by no means confined to this group of animals. Perhaps we should scarcely expect to find this kind of mimicry in such a comparatively highly organized a creature as a fish ; yet there is a group of fishes, familiar to those who have kept aquaria, under the name of sea-horses, in which it is exhibited in its full perfection. The ordinary sea-horse attaches itself to a sea-weed or some other object by curling its tail tightly round it ; and all these fishes have the habit of anchoring themselves by their tails in some way or another. In all of them the hard, horny body is furnished with a number of prominent ridges and spines ; but in one, belonging to a peculiar group from the Australian seas, these spines attain an enormous development, many of them being prolonged into irregular filaments or streamers of skin, which are especially developed throughout the long and slender tail. As these streamers float in the water, they so exactly resemble, both in colour and shape, the particular kind of sea-weed to which these fishes are in the habit of attaching themselves, that the whole creature seems but part and parcel of the fucus ; so that when on the sea-bottom it must be impossible for any carnivorous rover to distinguish between the animal and the vegetable.

One more instance of this kind of mimicry, and we must close this portion of our subject. This example is taken from the mammalian, or highest class of animals, and, although not such a perfect imitation of form as those we have already mentioned, is very remarkable as occurring so high up in the animal kingdom. Most of our readers are probably acquainted, at least by name, with those lowly South American mammals known as sloths. These animals are inhabitants of the great forest regions of that continent, and are of a sombre greyish colour, very like that of the gnarled and lichen-clad boughs from beneath which they are wont to hang back-downwards. Not only, however, is their general colour like that of a lichen-covered branch, but their coarse grey hairs actually develop a growth of lichens upon themselves to complete the resemblance to their surroundings. It is, indeed, clear that the long grey coat of the sloths has been produced for the sole purpose of this protective mimicry, for when this is removed there is found beneath an under-coat of softer fur marked by yellow and black stripes, which may be pretty confidently regarded as the original coloration of these animals.

We have now to consider animate, or true mimicry, in which one animal imitates the form, and generally the habits, of another in order to participate in the immunity from foes enjoyed by the latter, owing either to the possession of some formidable weapon, or to its unpalatable nature as food. In all cases of this kind of mimicry it is essential that the mimicked animal should be numerically far more abundant than the mimicker, as otherwise predatory creatures would soon learn that the innocuous and palatal animal was as likely to be captured as the harmful one. Undoubted cases of true mimicry are most common among insects, and it is to these alone that our observations will be confined. We may also observe that mimicking insects, as a rule, mimick other insects, although it has been considered that some large caterpillars mimick snakes, and certain moths certainly imitate birds.

We will first refer to some excellent and well-marked instances of mimicry which occur among insects of our own country. Most of us are probably familiar with those large hairy brown flies which may be seen in autumn creeping in a sleepy sort of manner about the windows of houses, and are commonly known as drone flies, and scientifically as *Eristalis*. These insects, although true flies, with only a single pair of wings, are so like bees (in which, it need scarcely be said, there are two pairs of wings) that it is very difficult to persuade some persons that they are not really members of that group of insects. Their resemblance to the latter is increased by their similar habits, more especially their bee-like buzz; and there is no doubt whatever but that they are mistaken by birds for bees, and thereby enjoy an immunity not granted to ordinary flies. An allied kind of fly (*Volucella*) goes even farther than this, and actually deceives the humble-bees themselves, which it closely resembles both in form and coloration. The object of this mimicry is to enable these flies to pass freely in and out of the nests of humble-bees in order to deposit their eggs, which eventually hatch into larvæ whose food are the grubs of the bees. Again, the gaudy flies marked with bold bands of black and yellow which are so common on fine summer days in gardens, and are known as wasp-flies (*Syrphus*), take their name from their resemblance to wasps, which in some species is so close as to make it difficult to convince people that they are not really wasps.

Less common than the above-mentioned flies are the beautiful British insects known as clear-winged hawk-moths. Some of these, named hornet clear-wings (*Sphecia*), so exactly resemble large wasps or hornets that they would deceive nine persons out of ten who are not entomologists. Moreover, they have exactly the same habits as hornets, and when caught will actually curl up their bodies in a wasp-like manner as if about to sting, although they are perfectly harmless. Less complete is the resemblance of other clear-wings—hence known as bee-clear-wings—to humble-bees. These insects, as has been well observed, are, however, very important, as proving that their mimicry is an acquired character, since when they first emerge from the chrysalis their wings are thinly covered with the well-known minute scales characteristic of ordinary moths, these scales soon falling off and leaving the wings perfectly transparent. This indicates that the ancestors of the clear-wings had wings like other moths.

In all the foregoing instances the mimicking insects

imitate various members of the Hymenopterous order ; but we have now to notice a case where a moth imitates a bird so completely as to deceive even the best observers when the two creatures are on the wing together. The moths in which this kind of mimicry occurs take their name of humming-bird hawk-moths from this very circumstance, and are represented by a species not very uncommon in some parts of our own country. So close is the resemblance between these moths and humming-birds that Mr. Bates tells us that, when on the Amazons, he has actually shot specimens of the former in mistake for the latter ; and the natives of these regions are firmly convinced that both are of the same species. The extended proboscis of the moth does duty for the slender beak of the bird, while the end of the body of the former is expanded into a kind of brush which imitates the tail of the bird. Humming-birds are, of course, not seized as prey by insectivorous birds, and hence the moths escape their natural enemies from their resemblance to the humming-birds. In our own country, where there are no humming-birds, it is somewhat difficult to see what advantage its bird-like form is to the humming-bird hawk-moth, and possibly its comparative rarity may be due to the absence of the birds it mimicks.

We come now to those very remarkable cases of mimicry, as exemplified among the butterflies, where one species mimicks one or even more members of the same order, owing to the immunity of the latter from the attacks of birds on account of their unpalatable taste. That the mimicked butterflies are protected by their unpleasant taste has been amply proved by their being offered over and over again to birds, by whom they are as invariably rejected. Their immunity from attack is further proved by their slow flight, and by the bright colouring of the under sides of their wings, so that they have no means of concealing themselves. Most of these mimicked butterflies are found in tropical and subtropical regions, and belong to the great families known as *Danidæ* and *Heliconiidæ*. In America these butterflies are usually mimicked by various species of the family of "whites" (*Pieridæ*), which, as we all know in the case of our common cabbage-butterfly, are eagerly sought by birds; and the difference of the mimicking species from an ordinary "white" by the assumption of the bright colours of the Danaids is so great that nobody but an entomologist would imagine for a moment that it even belonged to the same family. It is, moreover, curious that there is one instance where two species of Heliconids inhabiting adjacent regions are respectively mimicked by two varieties of one and the same species of "white."

Stranger even than this is, however, the case of certain South African swallow-tailed butterflies. In this group, as a rule, both sexes are alike, and furnished with the characteristic "tails"; but in one South African species the females entirely lose these appendages, and alter their coloration and the form of their wings so as to mimick not only one, but actually three distinct species of Danaids. Here, then, we have an instance in which a single species of butterfly exists under four totally distinct forms; viz., the typical swallow-tailed male, and the three varieties of tailless females respectively mimicking the three Danaids. No one would have the faintest idea that the three females belonged to the same family, let alone to the same genus and species, as the male; while the three varieties of the female would be assigned without hesitation to as many distinct species. That female butterflies are more often protected by mimicry than the males is a fact which may probably be explained by their extreme importance to the race, and also from the circumstance that when heavily laden with eggs they are more likely to fall a prey to birds than are the lighter males.

A great deal more might be said on the subject of mimicry in butterflies, but we must pass on to our last instance of this feature, which is, perhaps, the most peculiar of all. In this case the mimicked insect belongs to that peculiar group of ants which have the curious habit of carrying in their mouth a leaf which extends backwards over their bodies, and apparently acts as a kind of shade. Now, in British Guiana, there is an insect allied to the cicadas and other bugs, in which the leaf borne by these ants is represented by the thin and laterally flattened body of the creature, which is so compressed that it does not exceed a leaf in thickness, while its jagged upper border simulates well enough the irregular contour of the leaf carried by the ants, of which the borders are generally gnawed by the bearers.   Although the legs and lower part of the body of this most curious insect are reddish in colour, the leaf-like upper part of the body has assumed a green hue exactly resembling that of the ant-borne leaves.   In a drove of cooshie ants, as the leaf-bearers are called, the mimicking insect is distinguishable solely by its somewhat inferior size ; this difference is not, however, sufficiently great to attract the attention of birds, which have learnt by experience that the cooshies are by no means palatable morsels.

It would be beyond the scope of the present chapter to enter upon the difficult question of the means whereby these mimetic resemblances, whether to animate or inanimate objects, have been produced ; but sufficient has been said to show that an amount of interest lies in the subject, and those whose interest has thus been aroused may perhaps have the good fortune to discover new and unsuspected instances of one or other of these types of protective resemblances.

CHAPTER XVIII.

SEA-URCHINS.

PROBABLY most visitors to the seaside are more or less familiar with the shells of those marine creatures commonly known as sea-urchins, or sea-eggs, this acquaintance being usually due either to finding them cast up on sandy beaches, or to seeing them offered for sale by the vendors of curiosities and natural history objects. In many instances it is probable that the acquaintance ends here, although in others these objects may have been submitted to a fuller examination; but, in any case, we venture to say it is comparatively few who have studied them with the care and attention that their beauty of form and peculiar structure demands, and still fewer who know anything about their history in past times. Those, however, who care to take up the subject will find it one of more than usual interest, and we accordingly propose in this chapter to place before the reader some of the leading features and peculiarities of these creatures, which will form stepping-stones for those inclined to proceed further with their study.

To begin with, sea-urchins take their name from the array of movable spines with which the shell is covered during life, and which suggest comparison with those of the true urchin, or hedgehog. The shell, as it is commonly called, will, indeed, be the portion of the animal to which alone our attention will be directed, since it is only this part that is capable of preservation in a fossil state. We must, however, state at the outset that, as this so-called shell does not by any means correspond in structure with the shell of a mollusc, it is found more convenient to give it a different name, and the term *test* has accordingly been selected. Moreover, since the name sea-urchin is a somewhat long one, we may conveniently abbreviate it to urchin, unless we prefer to use the more technical term Echinoid.

There is great variety in the form of the hard calcareous

test of the urchins, which varies from a shape somewhat resembling a flattened orange to a heart-shape, or even to a thin disc-like plate. The ordinary urchins shaped somewhat like an orange are, however, those best adapted for gaining a general idea of the structure of the group, and we shall accordingly commence with them.

If, then, we examine such a test, we shall find that it has an aperture at each of the two poles; while it is divided into a series of meridional areas, each composed of a number of separate oblong calcareous plates, fitting accurately with one another at their edges, where they are united by a thin membrane. The upper surface of such a test is shown in Fig. 57, from which it will be seen that the upper or apical pole, as it may be conveniently called, consists of five somewhat heart - shaped plates surrounding a small circular orifice which is closed with membrane in the living state. This central orifice is the *vent*, while the five plates forming a star surrounding it, which generally fall out in the dry state, constitute the *apical disc*. At the opposite or lower pole, we shall find a much larger aperture, which forms the creature's mouth; and we may observe that just inside this aperture of the test in the complete animal there will be found a very complicated calcareous masticating apparatus known as *Aristotle's lantern*. Putting aside for the present the apical disc, we may devote somewhat fuller attention to the main body of the test, technically termed the *corona*. As we have said,

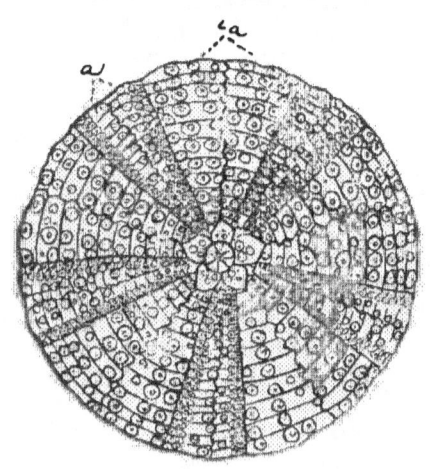

FIG. 57.—Upper Surface of the Test of the Common Sea-Urchin. *a*, ambulacral areas; *i a*, intermediate areas.

this consists of a series of meridional areas composed of numerous small plates ; and we shall find that these areas form ten alternating series, each of which consists of two meridional rows of the aforesaid plates. The line of division between the two rows in each area is well marked, although the divisions between any two areas are much less distinct. It will further be seen that, while in one series of areas (*i a*) the plates are much wider than in the other series (*a*), they are also somewhat deeper. Now, if we were to look from the inside of the test towards the light, we should find that the outer half of each plate in the narrower areas has several minute perforations ; and it is through these minute perforations that, during life, the animal protrudes the curious tube-like feet, by the sucker-like action of which an urchin is enabled to climb up the glass wall of an aquarium. Since these narrower areas are connected with the function of progression, they are appropriately termed the *ambulacral areas*, and the larger intervening spaces are accordingly called the *intermediate areas*. It is almost superfluous to add that the whole test of an urchin thus consists of five ambulacral and five intermediate areas, which between them comprise twenty separate rows of plates, each running continuously from one polar aperture to the other. During life, each plate is separated from its neighbour by a thin membrane, and the test increases in size both by additions to the edge of each plate and also by the interpolation of fresh equatorial zones of plates between the upper edge of the corona and the apical disc. The spines of the urchins are movably attached to the knobs with which the test is covered, and vary much in form and size, although the limits of this chapter do not admit of further reference to them.

Reverting to Fig. 57, we may also see that the ambulacral areas of the common urchin form a five-rayed star, on the test of which, in the position of the figure, three rays are turned away from the spectator and two towards him. Now this radiate arrangement at once forcibly reminds us of a star-fish, and any person who has ever handled those creatures when alive will be aware that from their under surface they can protrude tube-like sucking feet, precisely similar to those we have referred to as existing in the urchins. Both these points of resemblance are, indeed, indicative of an intimate relationship between urchins and star-fish. And, as a matter of fact, the ambulacral areas of the one represent the five rays of the other; the intermediate areas of the urchins being an addition to the structure of the star-fish. At first sight it looks, indeed, as if these animals were really symmetrically radiate, but we shall show later on that this is not the case, and that they are, in truth, bilaterally symmetrical like the higher animals, although this bilateral symmetry has been more or less thoroughly masked by the radiate arrangement of the parts.

Urchins and star-fishes are, however, not the only members of the group of Echinoderms; since this also comprises the beautiful stone-lilies or crinoids,* the joints of the stems of which form the well-known "St. Cuthbert's beads," of the Whitby lias, while the so-called entrochal marble, so often employed for chimney-pieces and other decorative architecture, is almost entirely composed of these stems. There are, moreover, certain entirely extinct types, such as the Cystoids and Blastoids, of which more anon.

Before proceeding to trace the modifications which the test of the urchins undergoes in different members of the group, it may be observed that in all urchins, whether recent or fossil, the number of meridional areas is invariably ten; while in every existing kind of urchin, no matter what be its size or shape, the number of rows of plates composing such areas never departs from twenty. As soon, however, as we reach the strata lying below the chalk and gault, known as the lower greensand, which constitute the lower part of the Cretaceous system, we find an urchin which departs somewhat in the last-named respect from the existing type. A side view of the test of this species is given in Fig. 58, from which it will be seen that while at the apical pole the number of meridional rows of plates is the normal twenty, as we approach the

---

* See page 157.

equator the series of meridional rows in each intermediate area is increased to four, which continue to the basal pole. Going still further back to the trias, we find another genus of urchins (*Tiarechinus*) with an increase in the number of rows of plates in the intermediate areas above the lower pole. Moreover, some of the urchins of the Jurassic rocks differ from the ordinary type, in that the plates of the test overlap one another instead of joining by their edges.

These departures from the normal form in some of the Secondary urchins suggest that, if we were to go back to the Palæozoic epoch, we should find still more marked differences from living types. Such, indeed, is actually the case,

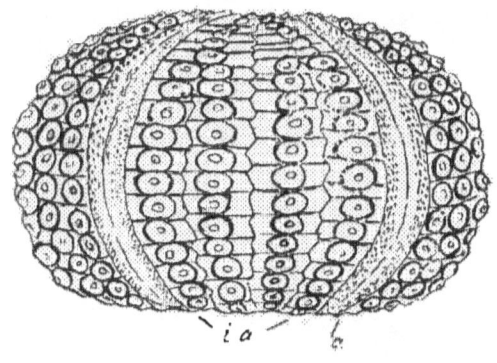

Fig. 58.—Side view of the Test of a Secondary Sea - Urchin (*Tetracidaris*). *a*, ambulacral areas; *i a*, intermediate areas.

and we find that all the Palæozoic urchins differ from existing ones in the number of meridional rows of plates, while very frequently these plates overlap one another. In the specimen represented in Fig. 59 it will be seen that while the ambulacral areas are normal, the intermediate areas have now no less than five meridional rows of plates; while in yet another form (*Melonites*) the ambulacral as well as the intermediate rows are increased in number, the former varying from seven to eight, and the latter from eight to fourteen. With one exception, however, all the Palæozoic urchins agree with the common species in having the vent situated at the apical, and the mouth at the basal pole, this mouth always having the "Aristotle's lantern."

Now if we proceed still further back till we reach the lower part of the Palæozoic epoch, such as the Cambrian

and lower Silurian, we shall find a totally extinct group
of Echinoderms known as Cystoids. The hard parts of
these creatures consist of a nearly globular test, usually
supported on a short stalk, and composed of a number of
polygonal plates, having no definite meridional arrangement, but traversed by five, or fewer, irregular ambulacral
grooves radiating from the mouth. And it will be obvious
that, in their large number of meridional rows of plates,
the Palæozoic urchins present a much closer resemblance
to these extinct Cystoids than is offered by their existing
representatives. Indeed, if we be-
lieve in the derivation of one form of
animal from another, it seems pretty
evident that starting from the Cystoids
—the oldest known Echinoderms—we
can pass readily into the Palæozoic
urchins, from which we are conducted
by the above-mentioned intermediate
Secondary forms to species of the type
of the common urchin of the present
day. What particular advantage the
modern urchins have gained by the
reduction of their meridional rows to
twenty is, however, not very easy to
determine; although this reduction
has probably conduced to greater
compactness and strength in the
structure of their test, which may
alone have been a sufficient improvement on the older types.

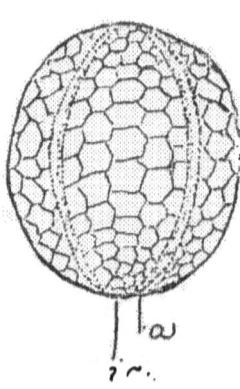

Fig. 59.—Side view
of the Test of a
Palæozoic Sea-Urchin
(*Palæoechinus*).   *a*,
ambulacral areas; *i a*,
intermediate areas.

The transition from the Palæozoic to the modern forms
does not, however, by any means exhaust the modifications which the urchins have undergone with the march
of time. Reverting once more to the common urchin
(Fig. 57), it should be mentioned that this type, in which
the test is radiately symmetrical and the vent and mouth
are polar, constitutes the group of the so-called regular
urchins. Although this type is still well represented, yet
a large number of the urchins of the Secondary, Tertiary,
and recent periods have departed very considerably from

this simple form, to assume a more or less decided heart-shape, with one or both apertures of the test becoming eccentric, and with a frequent tendency for the perforated portions of the ambulacral areas to become restricted to the central part of the upper surface, where they form a flower-like pattern (Fig. 60). Such types constitute the group of irregular urchins, which, on the doctrine of evolution, may be safely regarded as derived from the regular group. Moreover, not satisfied with the assumption of these new shapes, the irregular urchins appear to have considered the "Aristotle's lantern," which had served their ancestors as a masticating organ for countless ages, only a useless in-cumbrance, and, accord-ingly, the more advanced "radicals" among them have totally discarded this piece of apparatus, and, indeed, appear to get on equally well without it.

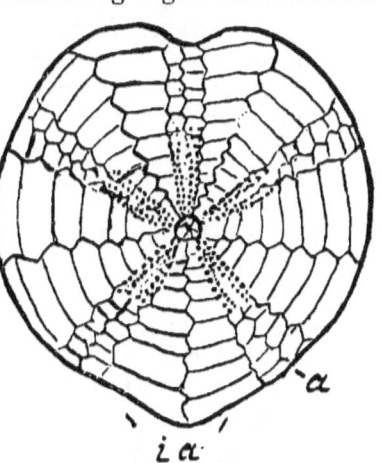

The forms connecting the more aberrant mem-bers of the irregular with the regular group of urchins are both numerous and varied. Among them we may refer to the com-mon little helmet urchin (*Echinoconus*) of our chalk. The test of this species is about an inch in height, and forms a tall

Fig. 60.—Upper surface of the Test of a Cretaceous Heart-Urchin (*Micraster*). *a*, ambulacral areas; *i a*, intermediate areas.

cone, which is not quite radiately symmetrical. The mouth, indeed, retains its usual position at the basal pole, while the perforated areas extend from pole to pole. In place, however, of the vent being situated in the centre of the apical disc, it has been transferred to the margin of the lower surface, in the middle of one of the intermediate areas. It is thus placed opposite to an ambulacral area corresponding to the one directed upwards in Fig. 57;

and we thus, for the first time, find a regular bilateral
symmetry fully established.　　In the particular species to
which we have referred (as in all the under-mentioned
forms), the "Aristotle's lantern" has been lost, although
it is retained in other members of the same group.　One
specimen of a helmet urchin has been discovered showing
the rare abnormality of having but four ambulacral areas.

A further step is shown by the so-called sugar-loaf
urchin (*Ananchytes*) of the English chalk, the tall silicified
tests of which are so commonly found in the gravel derived
from the denudation of the chalk.　In this species, which
is considerably larger and more ovoid than the helmet
urchin, the elevated form of the test is still retained, but
the contour of its under surface has become more decidedly
heart-shaped.　Moreover, the vent has become shifted
completely on to the lower surface; while the mouth,
although remaining (as is invariably the case) on the
under surface, has travelled away from its original central
position so as to be placed in the ambulacral area corresponding to the one directed upwards in Fig. 57.　Here,
therefore, we have the bilateral symmetry still more
distinctly marked.　Another advance shown in this urchin
is the circumstance that the perforated portions of the
ambulacral areas, instead of extending on to the lower
surface of the test, stop short at the edge.　Some members
of the same family are still more peculiar in that the apical
disc is greatly elongated from front to back, in consequence
of which the five ambulacral areas do not all meet one
another at the summit of the test: those corresponding to
the three turned away from the spectator in Fig. 57
meeting near the middle of the test; while the remaining
two are brought together in the opposite part of the test
corresponding to the lower portion of the same figure.

A third common chalk species, the heart-urchin, or
fairy heart of the quarrymen (*Micraster*), affords a good
example of what may be regarded as the extreme modification of structure developed in the group.　The upper
surface of one of these urchins is represented in Fig. 60,
from which it will be seen that the contour is regularly
heart-shaped, and the whole test much depressed.　The

perforated portions of the ambulacral areas are now restricted merely to the central region of the upper surface, one of these areas (directed upwards in the figure) forming a shallower groove, and being otherwise markedly different from all the others. In such an urchin the bilateral symmetry is very strongly marked indeed, and, since the upper border of the figure represents the anterior, and the lower the posterior extremity of the animal, we may compare the three anterior ambulacral areas to the head and arms of a quadruped whose hind legs will be represented by the two posterior ambulacra. In regard to the position of the two orifices of the test in this species, the vent is situated on the flattened posterior surface, near its junction with the upper surface, while the mouth occupies a position nearly midway between the centre and the anterior border of the lower surface, at the commencement of the groove formed by the anterior ambulacral area. The mouth is peculiar in that it does not open directly on the surface, in the usual manner, but has a projecting lip by means of which its aperture assumes a forward direction. The common purple heart-urchin (*Spatangus*) of our present seas is a larger representative of this group, presenting the same general type of structure.

We thus see how gradual is the passage from a species of the type of the common urchin to that of the heart-urchin, widely different as are these two from one another. We might, indeed, proceed further with our investigations, and show how certain of the irregular urchins have become so flattened as to assume the form of thin plates, which in some instances are deeply notched at their periphery. And we might also investigate the variation of form and size displayed by the spines of the different groups. Enough has, however, been written for our present object, which has been to show the amount of interest that attaches to the investigation of the lines of modification on which development has proceeded among the sea-urchins. This has shown how a regular progressive advance in one particular direction has taken place from the earlier to the later members of the group ; and we thus have another

excellent instance testifying in favour of the doctrine of
evolution. This brief sketch may possibly give additional
pleasure to a sea-side sojourn, by inducing some of our
readers to direct their attention, first of all, to the recent
sea-urchins, after which they will scarcely fail to extend
their investigations to the fossil species so abundantly
distributed through our rocks ; and, we will venture to add,
that if they do so their interest cannot fail to be aroused.

CHAPTER XIX.

NUMMULITES AND MOUNTAINS.

BOTH the proverb "as old as the hills," and the phrase the "everlasting hills," are but the expression of the natural tendency of the human mind to regard all hills and mountains as the most lasting and ancient objects with which we are familiar. As, however, is so often the case, science steps in and tells us that, although the proverb and the phrase are true enough as regards human experience, yet that when we go back and study the origin of things, as revealed by geology, we find that many hills and mountains—and more especially the highest of them—are actually among the very newest features in the physiognomy of the earth, and that the expression "as old as the plains" would, in many instances, be a far truer simile than the one current.

If, indeed, we reflect for a moment, we shall be convinced that the highest mountains of the globe—always excepting volcanoes, which have the power of renewing their height—must necessarily, as a general rule, be younger than many of those of lower height, or even than the plains from which they rise. Thus rain, snow, frost, and rivers are perpetually tending to wear down and wash away all the higher points of the earth's surface, and to carry the *débris* to the valleys below. Consequently, we are led to conclude that, *primâ facie*, the higher a mountain range is, the less time it has been subject to this washing-away process, and, therefore, the younger it is as regards relative age. It might, indeed, be objected to this that the mountains that are now the highest have always been the highest; and that at the beginning of all things their original height as much exceeded their present height as the latter does that of the smaller ranges. In this view, however, their original heights would have had to be so stupendous as to be almost inconceivable, and

o 2

probably much greater than is compatible with the physical
conditions of the globe. This hypothesis may, therefore,
be dismissed as untenable; more especially as there is
direct evidence of a totally different kind, which is
conclusive as to the truth of the alternative view.

This evidence is afforded by fossils, and more especially
by a particular kind of fossil, which, from its abundance
and the restricted geological epoch in which it is found in
any quantities, is of more than usual value in inquiries
of the present nature.

If any of our readers have ever examined the Tertiary
clays and sands of Barton, in Hampshire, or of Brackle-
sham, on the Sussex coast, they will probably have met
with numerous disc-like objects, the larger of which are
somewhat more than an inch in diameter, while the
smallest are scarcely bigger than a pin's head. When
split or cut, these objects are found to contain a number
of minute chambers, separated from one another by thin
walls arranged in the form of a spiral, as shown in the
figure. Technically they are known as nummulites, and
belong to the very lowest division of the animal kingdom—
lower even than the sponges, which some people cannot
be persuaded to believe are animals at all. Now these
nummulites are exceedingly interesting to those who study
the growth and formation of mountain ranges for the
reason that they occur, in any quantity and of large size,
only through the greater portion of the Eocene or lowest
division of the Tertiary (latest) geological period, although
not reaching down to the London clay; and also because
they were very widely distributed in the seas which then
covered a large part of our existing continents. If, then,
we should find rocks containing great numbers of large
nummulites on the flanks or tops of a mountain range, we
should be assured that such range was younger than the
Eocene period, at which date its component rocks were
being formed as mud at the bottom of the sea.

Now, although in England the aforesaid nummulites
only occur in soft beds of clay and sand in the low cliffs
of the southern coast, when we cross to the Continent we
find them forming the greater part of a massive limestone.

known as the nummulitic limestone. This very characteristic rock is more massive and more widely spread
than any other Tertiary deposit, and, in its thickness and

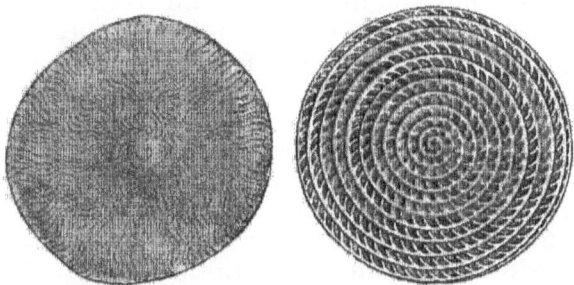

Fig. 61.—A Nummulite, viewed from above, and horizontally bisected.

identity of structure over large areas, recalls the mountain
limestone of the Palæozoic epoch. It is, indeed, absolutely one mass of nummulites, of which sections are
displayed on every fractured surface ; and it was probably
an open sea deposit, which must have taken incalculable
ages for its formation. It occurs in southern Europe in
both the Alps and Pyrenees, attaining a thickness of
several thousand feet in the former, and occurring at
elevations of more than 10,000 feet above the sea-level.
In the Pyrenees it forms a beautiful white crystalline
marble. On the south of the Mediterranean, nummulitic
limestone is found again in the mountains of Algeria
and Morocco. In eastern Europe it reappears in the
Carpathians, and thence may be traced into the Caucasus
and Asia Minor. All travellers to India are familiar with
the Mokattam range of bare mountains on the western
shore of the upper part of the Red Sea, which are likewise
almost entirely composed of this same limestone. It is,
indeed, a common belief among the Egyptian peasantry
that the larger disc-like nummulites are lentils left by the
builders of the pyramids, and subsequently turned into
stone. From the Caucasus and Asia Minor the nummulitic limestone may be followed into Persia, Baluchistan,
Sind, the Punjab, and so into the Himalaya. Thence

it continues into Assam and Burma, and reappears in the Andaman and Nicobar Islands in the Bay of Bengal.

It is, however, in the inner Himalaya that the occurrence of nummulitic limestones and certain overlying Tertiary rocks is of more especial interest, since it is there that they attain a greater elevation than in any other part of the world. It is in the upper Indus Valley, in the neighbourhood of Leh in western Tibet, that these nummulitic rocks occur; running some distance down the Indus to the west of Leh, and to the eastward of that town extending into Chinese territory. There is good evidence to show that the arm of the sea in which these nummulitic rocks were deposited communicated with the ocean to the eastward in the Bay of Bengal, instead of following the course of the Indus in a westerly direction to the Arabian Sea. Moreover, in some parts of this area the rocks which overlie, and are, therefore, newer than the nummulitic limestone, are raised to the stupendous elevation of more than 21,000 feet above the sea-level.

We have, therefore, before us decisive evidence to show that those parts of the earth's surface which at the present day form some of the highest peaks in the Himalaya were, at the period when the London clay was deposited, below the level of the sea; and consequently that the elevation of that part of the Himalaya has taken place entirely since that epoch, during a period when the physical features of England have altered only to a comparatively slight degree. There is, moreover, equally conclusive evidence to show that the elevation of the Himalaya was not completed until a much later epoch of the earth's history, since on the southern flanks of this mighty range we find beds of sandstone containing remains of mammals which lived during the Pliocene, or later Tertiary epoch, raised to a height of several thousand feet above the sea-level.

The elevation of the Indus Valley in the heart of the Himalaya could not, therefore, have commenced until the Miocene, or middle Tertiary epoch, while that of the outer Himalayan ranges could not have been completed till far into the Pliocene period, and, for all we know to the contrary, may still be in progress. Not only so, but the same evidence likewise tells that the Alps, Pyrenees, Carpathians, the Caucasus, and the Egyptian Mokattam range, as well as the mountains of Algeria, have all attained their present elevation since the latter part of the Eocene period, when at least a considerable portion of their area was submerged. And we accordingly learn that many of the most striking physical features of the Old World are of comparatively modern origin.

When, however, we turn to mountains like those of the Lake District and Wales, which only attain moderate elevations, and in which the rocks belong solely to the Palæozoic, or oldest geological epoch, it is evident that we have to do with elevations of an extremely remote date. There is, indeed, satisfactory proof that these old mountains were once vastly higher than they are at present; their diminished altitude being due to the long ages during which they have been subjected to the wear and tear of the elements. To such mountains the proverb to which we have already alluded is, therefore, strictly applicable; but in a geological sense the phrase "everlasting hills" can be applied neither to the oldest nor the youngest mountains.

CHAPTER XX.

A LUMP OF CHALK AND ITS LESSONS.

PROBABLY all Englishmen—certainly all those dwelling in the eastern and south-eastern counties—are familiar with the pure white rock which we call, from the Latin *creta*, chalk. It is indeed this very familiarity which breeds the proverbial contempt, and causes us to take but scant or little notice of what is really a very beautiful substance in itself, altogether apart from the interest with which it is invested from a geological point of view. If chalk were very rare instead of being exceedingly abundant, there is little doubt that it would be reckoned as a beautiful substance, worthy to stand as the best example of a pure white mineral alongside of virgin sulphur as the finest example of a yellow one. If, moreover, chalk had happened to have undergone the action of intense heat under equally intense pressure, it would assuredly have produced an even finer and purer statuary marble than that of Carrara, and might thus have been one of the most valuable of rocks.

A complaint may not unfrequently be heard among those more or less deeply interested in geological science who happen to dwell in a chalk country, that the very sameness of the chalk formation throughout England prevents them from finding any interest in the geology of their own districts, and thus leads them to regret that their lot had not been cast in regions where a variety of rocks are to be met with. Although there is a considerable amount of truth in this complaint, yet if rightly studied the chalk is so peculiar and unique a formation as rather to embarrass us with the number of considerations and problems to which it gives rise, than to be deficient in interest.

Examining a lump of the pure white chalk of many parts of England, such as that of Dover, we find that it consists, both under the naked eye and an ordinary lens, of an exceedingly fine-grained homogeneous soft substance, adhering strongly when applied to the tongue, and leaving a white streak when rubbed on other substances. If treated with vinegar, or any other acid, it will effervesce strongly with the liberation of the gas commonly known as choke-damp, or carbonic acid, while the base unites with the new acid to form a fresh compound of lime. The lime may be obtained in a pure condition by burning the chalk, as in a lime-kiln, when the carbonic acid is likewise given off; and we thus learn that chalk consists of carbonate of lime. As a rule, when we examine a chalk-cliff we shall find that the chalk, although stained here and there with iron, is identical in structure throughout great thicknesses, and that it shows nowhere any signs of crystallization. Occasionally, however, as at Corfe Castle, near Swanage, in Dorsetshire, we shall find that the chalk has become so hard as to leave no distinct streak when rubbed lightly on other substances; while its cracks and fissures are filled with translucent crystals of white spar—the calc-spar, or calcite of mineralogists. Here, then, we have the chalk so hardened, probably by the effects of subterranean heat, as to form what is popularly called a limestone; while a farther step would have converted it into actual marble. The geologist would indeed apply the name limestone to chalk, ordinary limestone, and marble indifferently; but since the popular usage is different, it is well to be assured that all three are but various modifications of one and the same substance. In the north of Ireland the basalt of the Giant's Causeway has converted the chalk still more completely into a hard limestone.

Chalk, then, may be defined as a fine-grained, white, non-crystalline, soft limestone. This, however, by no means exhausts the subject of its composition. Thus if we take a piece of chalk and wash it carefully in water with a hard brush so as to reduce it to a state of mud, and examine the portion which falls to the bottom of the vessel under a microscope, we shall find that this is very largely made up of various shell-like substances. Many of these are minute fragments of what may have been real shells, while others are portions of the spines of sea-urchins, and others, again, are the flinty spicules of sponges. By far the larger proportion consists, however, of perfect objects of extremely minute size, mainly belonging to that lovely group of animals known as foraminifera, or shortly, "forams." Of these beautiful little shells some are coiled in a manner recalling the shell of the nautilus, while others consist of globular masses arranged either in a coil or in a straight line, the globules gradually increasing in size from the summit to the mouth of the shell. In all cases, however, the walls of these shells are perforated by the inconceivably minute apertures from which the forams take their name, and through which, when alive, the creatures protruded delicate threads of the jelly-like protoplasm of which their soft parts are composed. Truly marvellous in beauty are these forams, although pages of description can give but a faint idea of them, and the student should see them for himself under a microscope. So numerous, however, are these forams, and other equally minute organisms in the white chalk, that they frequently compose half its substance, while it is stated that in some rare cases they may even rise to as much as ninety per cent.

We have now, therefore, to add to our definition of chalk that it is largely composed of the shells of the minute animals known as forams, together with those of other allied creatures, and so many accordingly speak of it as a limestone which is evidently, to a large extent, of organic origin. Moreover, as these forams are more or less closely allied to species inhabiting the ocean at the present day, we should be justified from this evidence alone in regarding the chalk as a formation of marine origin. This origin is, however, equally well proved by the larger fossils, such as the shells of sea-urchins, scallops, oysters, &c., commonly occurring in the chalk ; while, in addition to this, the extreme purity and thickness of the formation would of itself be sufficient to demonstrate that the chalk is the result of long-continued deposition on the bottom of the sea.

Thus much for the composition of our lump of chalk as

examined in the laboratory, and we now turn, as all geologists worthy of the name should, to its occurrence in the field. If we look at one of the tall chalk cliffs of our southern coasts, as in the neighbourhood of Dover, we shall be first of all struck with the extreme homogeneity and purity of the whole formation from top to bottom, through a thickness which in this neighbourhood is close upon 1000 feet, and in Norfolk is more than 1100 feet. This similarity of composition throughout such a vast thickness is totally unlike what we are accustomed to observe in other rock-cliffs (although there is some approach to it in the blue mountain limestone of Derbyshire), where we generally find alternating bands composed of rocks differing both in colour and structure from one another, and we are thereby led at once to conclude that there must be something very peculiar connected with the deposition of the chalk. How was it that in the old sea there were not only no currents bringing loads of sand or clay to alternate with the pure white limestone, but, above all, that there was not a tinge of colouring matter to stain the virgin purity of the newly-forming chalk during those ages and ages of time, while drifted logs and fruits occur but rarely?

A closer inspection of a large thickness of chalk will, however, reveal the fact that there is not a complete similarity in the nature of the rock throughout the entire formation. Thus, whereas in places where nearly the whole formation is displayed we find throughout the uppermost 400 feet layers and nodules of flint are thickly distributed throughout the mass, generally forming more or less well-marked lines which indicate the original planes of the deposition of the rock, as we pass to a lower level the proportion of these flints becomes gradually less, till, after we have passed downwards through some 130 feet, they finally disappear, and are wanting throughout the whole of the lower part of the series. Moreover, in this lower chalk, or chalk without flints, we shall find, as we pass downwards, a gradual tendency to lose the pure white colour of the upper chalk, and to assume a buff or greyish tint, while in the very lowest beds we shall not fail to notice the appearance of a number of small grains of a greenish-coloured mineral. If, again, we try to dissolve this lower chalk in acid we shall find that as we descend in the series there is an ever-increasing quantity of an insoluble remnant, which would be shown by analysis to be of the nature of clay. Both these circumstances point to the conclusion that at the time the lower chalk was laid down the conditions were by no means so well adapted for the deposition of a pure carbonate of lime as was the case in the later time of the upper chalk with flints. What these conditions were we shall consider subsequently, but we have now to direct our attention to the area over which the white chalk extends.

In the north-west the farthest limits to which the white chalk extended are found near Belfast, where, as we have said, the rock has been converted into a hard limestone by the action of heat. Although we do not again meet with chalk till we reach the east and south of England, where it forms large portions of our coast from Dorsetshire to Yorkshire, yet it is probable that the chalk sea embraced the foot of the Welsh mountains, which formed an archipelago. From England the white chalk may be traced without any alteration in its character through the north of France, the south of Belgium, the eastern part of the Netherlands, and thence through Westphalia, Hanover, and Galicia, into Poland and Russia, where it reaches on the one side to the foot of the Urals, and on the other to the Crimea ; moreover, to the northward it occupies a considerable portion of Denmark and the southern extremity of Sweden. Although the white chalk is now only distributed over the surface of this region in larger or smaller patches, being sometimes covered up by newer (Tertiary) deposits, and in other places totally wanting, there is evidence that it once extended continuously over the whole. Moreover, the absence of any traces of the white chalk in the regions to the west and north of those mentioned, indicates that the present limits of the chalk in those directions mark approximately the boundaries of this cretaceous sea, this sea being probably cut off from free communication with the Atlantic by a barrier connecting western France with Cornwall and Ireland, and by another joining Scotland with Scandinavia.

The above area includes the whole of the white chalk; but when we trace this chalk southwards into Bohemia and Saxony we find that it has undergone a very remarkable change. Thus, although it contains the same fossils as to the northward, the rock itself, instead of being the pure white limestone to which we have been accustomed, consists of a series of massive sandstone about as unlike chalk as anything well could be. It is probable, indeed, that these cretaceous sandstones, as we may call them, were formed in a gulf on the southern coast of the white chalk sea, which was unfavourable to the deposition of chalk itself; and as these sandstones were undoubtedly deposited at the same time as the pure chalk, we thereby learn the very important geological lesson that similarity or dissimilarity in the mineralogical structure of a rock is a matter of very minor import indeed. We may illustrate this by reference to architecture. Thus, a Gothic church may be built either of sandstone, limestone, marble, or, for the matter of that, brick; but it will still be (exclusive of course of our so-called modern Gothic) absolutely characteristic of one particular period of European architecture. This Gothic style will be distinguished by certain peculiarities in the structure of its arches and pillars, as well as by the ornaments with which they are embellished. Just so in geology we have a chalk or cretaceous style, in which, although the rock itself may be either chalk or sandstone, or limestone, or slate, yet its architectural details—that is to say, its fossils—will be the same, not only throughout Europe, but within certain limitations of variation, over the whole world. This is one of the important lessons to be learnt by a comprehensive study of our white chalk.

The second great lesson taught by the white chalk is, however, of perhaps still more importance. We have seen that the white chalk was deposited in a sea cut off from free communication with the Atlantic to the west and north; and the range of the Ardennes which formed its shore in the south-west, together with the evidence of the

near neighbourhood of a coast afforded by the sandstones
of Saxony and Bohemia, indicates that this sea was a
mare clausum (in a geographical, not a political sense),
somewhat like the Mediterranean or the Black Sea. Now,
from the apparent similarity of chalk to the ooze forming
in the abyssal depths of the Atlantic and the other large
ocean basins, it was taught but a few years ago that the
chalk itself was deposited in an ocean of similar depth.
The *mare clausum* theory, however, is of itself a sufficient
obstacle to the acceptance of such a view, since it is
impossible to conceive that a sea of such small dimensions
could ever have had depths at all approaching those of the
Atlantic. The Atlantic theory, if we may so call it, of the
chalk was, however, at once and for ever dissipated by
the researches carried on during the voyage of the
"Challenger." Those researches showed that the so-
called abyssal deposits, instead of being very like the chalk,
were really very different. Even the ooze has not the
purity of the chalk; while the large areas of red clays
covering the ocean-basins have no analogy in the latter.
Moreover, it has been proved that the abyssal deposits are
laid down at a rate of almost inconceivable slowness—so
slowly indeed that even meteoric dust forms an appreciable
portion of the red clays; while the ear-bones of whales
and teeth of sharks that strew the ocean-floor have lain
there so long as to have become coated over with a thick
layer of manganese precipitated from the water of the
ocean. On the other hand, the remains of fishes and
other delicate organisms which occur so beautifully preserved in the white chalk clearly indicate that its
deposition must have been comparatively rapid, and must
have taken place in a sea where there was abundance of
mineral matter either in suspension or solution. Again,
the fauna of the chalk, especially the sponges, is one
such as would be found in comparatively shallow seas,
and is quite unlike that of the Atlantic depths. Indeed, it is quite probable that the chalk sea may not
have exceeded some one or two thousand feet in depth.
The great difficulty in regard to the chalk is, indeed, to
explain its purity, and the very rare occurrence of drifted

materials found imbedded in it. The *mare clausum*, with no tides and perhaps but few large rivers flowing into it, and its shores largely composed of hard crystalline rocks like those of Scandinavia and the Ardennes, will, however, to a certain extent remove this difficulty. Even then, however, it is doubtful how sufficient material for the formation of the chalk could have been obtained; and accordingly one of our most eminent living geologists suggests that, in addition to its partially organic origin, chalk may have been largely formed by a chemical precipitate of carbonate of lime.

Be this as it may, the degradation of chalk from its former position as a supposed typical abyssal deposit has taught the great lesson that almost all the stratified rocks with which we are acquainted were laid down in comparatively shallow water, and consequently has led to the general acceptance of the grand doctrine of the permanence of continents and ocean-basins. By this, of course, it is not meant that the whole areas of several of our continents, such as Europe, have not been (as we know they have), many times over, beneath the sea. Indeed, what we have already said as to the extent of what we may call the cretaceous Mediterranean, shows that at a comparatively late period of geological history a large part of central Europe was sea. Neither does this doctrine forbid such changes in the present configuration of the earth as would be implied by a land connection between Africa and southern India. What, however, it does say, and that in the most emphatic manner, is that where continents now are there deposits have always been going on, and there land, of larger or smaller extent and of ever varying contour, has always been; while the great ocean-basins, like those of the Atlantic and Pacific, have existed as such since the globe emerged from its primeval chaos. This, then, is the second great lesson taught by a lump of chalk!

We have, however, by no means yet exhausted the interest connected with the subject of chalk. In the first place, the gradually increasing marly character of the lower chalk points to a condition when the sea was much less deep than at the period of the white chalk. If, indeed,

we go lower down in the rock series, we shall find the white
character of the chalk has completely disappeared when
we reach the underlying blue "gault" of Folkestone, which
implies the existence of currents or rivers largely charged
with mud. Still further back, we have the fresh-water clays
and sandstones of the Weald of Kent and Sussex ; and
we thus learn that at that period southern England was in
the condition of a large delta, after which there was a
gradual subsidence, culminating in the *mare clausum* of the
period of the white chalk. Then, again, we have seen
how the " architectural style " of a rock, as exemplified by
its fossils, is the one all-important point connected with
it; and the alteration of the English chalk into the
cretaceous sandstones of Saxony ought to have prepared
us for more extensive modifications of these rocks as we
proceed to regions still more remote from where they are
typically developed. If, then, we turn to a geological map
of Europe, we shall find a large area of its southern half
coloured in, as being formed of cretaceous rocks—that is,
rocks equivalent in point of age to the white chalk. The
description, or still better, an actual examination of these
rocks will show, however, that they have but little in
common with the white chalk. They consist, indeed, of
hard, compact, and often dark-coloured limestones, containing many fossils identical with those of our own chalk,
together with certain others of different types; thus
showing that we have entered an area where the conditions
of life were somewhat different from those obtaining in
the *mare clausum* of the white chalk. From the centre
and south of France these cretaceous limestones may be
traced across the Pyrenees into Spain, and so into North
Africa, while eastwards they extend across the Alps into
Switzerland, Italy, Bulgaria, Roumania, and thence along
the Mediterranean basin into Asia. That these rocks
stretch far into the heart of Asia is now well known, and
since rocks of somewhat similar type containing well-
known European cretaceous fossils are found in the inner
Himalayas, it seems highly probable that this southern
cretaceous sea connected the Mediterranean with the ·Bay
of Bengal. Whereas similar cretaceous fossils occur on

the east coast of India, in the neighbourhood of Madras, and since there are some very remarkable similarities between the fresh-water rocks of the peninsula of India and those of South Africa, while many animals are now common to those two countries, there are very strong reasons for considering that peninsular India (which was then cut off from the rest of Asia by the cretaceous sea) had a land connection with the Cape by way of Madagascar. We know indeed that this southern cretaceous sea communicated freely with the Atlantic, by what is now Spain and France, and we are thus led to conclude that there was formerly a direct sea communication between the Atlantic and the Bay of Bengal by way of central Asia. Europe and Asia then formed a northern continent separated by this cretaceous sea (of which the Mediterranean is the shrunken remnant) from a southern continent which included both Africa and India proper. Such is the wide interpretation given to the doctrine of the permanence of continents and ocean-basins.

The study of the European chalk, besides the two great lessons to which we have especially directed attention, has, therefore, proved to us the former existence of two great seas, in which the cretaceous rocks were deposited—the northern one, being a *mare clausum*, cut off from the Atlantic, in which was deposited the white chalk ; while the southern one, in which the hard, massive limestones of southern Europe were laid down, proved the connecting link between the Atlantic and the Indian Ocean, to which we have already alluded. We might pursue our subject further, and discuss the origin and nature of the flint and pyrites which are of such common occurrence in the chalk, or we might direct attention to the more valuable and much rarer phosphates which are sometimes contained in it. We might, again, discuss the peculiar characters of the cretaceous fauna, and show how that of the closed northern sea differed from that of the open southern ocean. We might do all this, and more ; but what has been written is sufficient to show the amount of interest and the many weighty problems connected even with a "lump of chalk."

P

CHAPTER XXI

A FLAKE OF FLINT AND ITS HISTORY.

IN the foregoing chapter it was stated that the upper
white chalk, so far as hand specimens are concerned, is a
nearly pure limestone. This, indeed, is a perfectly true
statement as regards such hand specimens; but when we
consider the upper chalk as a whole, we must not omit to
regard the numerous bands and nodules of flint with
which it is interstratified as a very important constituent
of the whole rock. We say a constituent of the whole
rock advisedly, because although the flint is now separated
from the white limestone which we call chalk in the form
of nodules and bands, yet there is evidence that it was
originally disseminated throughout the entire mass, and
that the upper chalk then formed a slightly siliceous
limestone. Probably everybody is more or less familiar
with flint as it occurs in a chalk-pit, or in the form of
gravel derived from the disintegration of chalk strata;
but it may be taken for granted that comparatively few
have ever seriously considered how the solid masses of
flint have originated in the soft chalk limestone. As this
is a subject of considerable interest, and one which has
given rise to much discussion, we propose to devote the
greater part of the present chapter to its consideration,
while we shall add some observations on the history of
flints after they have been removed from their native
chalk.

We shall assume, in the first place, that all our readers
are aware that flint is one of the manifold forms assumed
by that abundant constituent of the earth's crust technically
known as silica—the oxide of the element silicon. When
crystallized, silica occurs in the form of rock-crystal, or
quartz; but flint is one of the many non-crystalline, or
amorphous, developments of the mineral. It is generally
defined as a massive dark-coloured or black, semi-trans-
lucent, dull-looking variety of silica; which when pure

burns to an opaque white, and has what is termed a
conchoidal or shell-like fracture. The dark colour, it may
be added, is due to the presence of a small quantity of
organic matter, or carbon.

If we fracture a nodule of flint freshly taken from a
chalk-pit, we shall find that the thin edges of the sharp
flakes which are seen to be produced have a pale-brown
horn colour when viewed by transmitted light, and that
as the flake becomes thicker the colour gradually darkens

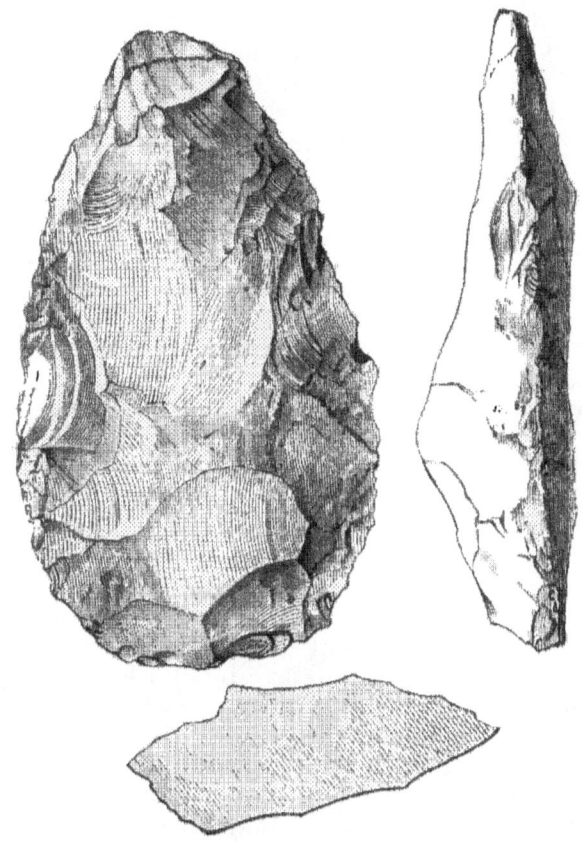

Fig. 62.—A Chipped Flint Implement, from Icklingham, half
natural size. (From Sir J. Evans' "Stone Implements.")

P 2

till it assumes the blue-black tint characteristic of the mass. If, however, our flake contains a portion of the external coat of the nodule, we shall see that for about a quarter of an inch in depth the outer layer is far less compact than the rest of the flint, and is of an opaque white colour. In other specimens, again, we shall observe that the colour, instead of the usual deep blue-black, is of a pale whitish-blue, frequently marked by more or less distinct bands. Needless to say that on trying to scratch our flint flake with a knife we shall signally fail, and if any result happens it will be that a thin film from the metal of the blade will be left on the stone at the point of contact.

Our flake will likewise exhibit in great perfection the characteristic conchoidal fracture of flint ; that is to say, its surfaces will be smooth and undulating, swelling here into a prominent convexity, and falling there into a deeper or shallower hollow. Frequently, moreover, there may be observed a number of small parallel wavy ridges on the fractured surface. If we submit the edges of the freshly-broken flake to a series of taps from our hammer, we shall find that a number of smaller flakes will be readily chipped off, each leaving a separate conchoidally-fractured surface on the original flake. It is this facility with which flint can be chipped, coupled with its hardness and the sharpness of its fractured edges, that induced our palæolithic ancestors to adopt it as the material for their various weapons and tools. The figure on the preceding page of one of these implements exhibits in great perfection the characteristic conchoidal fracture of flint.*

The extreme development of this peculiar and characteristic fracture is, however, exhibited when a large flat surface of flint is struck at right angles by a round-ended hammer. The hammer then comes in contact with a minute portion of the surface of the flint, which may be represented by a small circle, and as the flint is elastic, "this small circle," as Sir John Evans observes, "is

---

* We are indebted to Messrs. Longman and Green for this and the following figure.

driven slightly inwards into the body of the flint, and the result is that a circular fissure is produced between that

FIG. 63.—Artificial Cone of Flint. (From Sir J. Evans.)

part of the flint which is condensed for the moment by the blow, and that part which is left untouched. As each particle in the small circle on which the hammer impinges may be considered to rest on more than one other particle, it is evident that the circular fissure, as it descends into the body of the flint, will have a tendency to enlarge in diameter, so that the piece of flint it includes will be of conical form, the small circle struck by the hammer forming the slightly truncated apex." A little practice will enable anyone to make these flint cones with ease. The size of the cone, and the degree of steepness of its sides, vary with the nature of the flint, the weight and form of the hammer, and the force of the blow.

When examined with a lens or microscope, chalk-flint frequently exhibits a perfectly uniform structure throughout, without the least trace of the presence of any organic body. In other cases, however, traces of sponges, corals, shells, echinoderms, diatoms, &c., are more or less apparent in flint. Perhaps the most common of all these organisms are the mushroom-shaped sponges known as ventriculites; and if we examine a flint containing one of these sponges we shall frequently observe that there is a complete transition from portions of the perfectly preserved sponge to homogeneous flint, without the least trace of organic structure. In the case of echinoderms, we shall find that while in some cases the whole interior of the shell (or, as it is technically called, test) is filled with flint, the shell itself retaining its original calcareous structure, in other instances the original shell itself has been completely removed and replaced by flint. In some cases the original structure of the shell has been preserved in the flint, but more generally this has been completely lost, and the flint is structureless.

This replacement of a calcareous by a siliceous structure is an instance of pseudomorphism. Similar remarks will apply to the shells of molluscs, and likewise to corals. It is important to mention that the sponges found in flint were not originally of the horny nature of our bath-sponges, but were themselves composed of minute spicules and fibres of silica, like the so-called Venus's flower-basket of our modern seas.

With regard to the mode of occurrence of flint, we have first to mention that it is by no means confined to the chalk, but may occur in limestones of any age. In this country it is, however, more abundant and purer in the chalk than in any other formation, and may, indeed, be considered characteristic of the upper part of that formation.* Flints are found in the chalk either in the form of nodules or in thin continuous laminæ. The nodules are generally of very irregular shape, and may vary in size from a walnut to masses of a hundred-weight or so. As a rule they occur in strings at comparatively regular intervals in the chalk, generally conforming more or less closely to the original planes of bedding, and the individual nodules being sometimes at considerable distances apart, but at others closer together and more or less connected by long root-like pieces. On the other hand, the laminated or tabular flint may cut the bedding-planes of the chalk at any angle, and is often found in joints and fissures, which may emerge at the surface. As a rule, this tabular flint is devoid of organisms. Not unfrequently flints may be found which near their surfaces gradually become paler and paler in colour, and contain an increasing amount of calcareous matter, till they pass imperceptibly into hard siliceous chalk. Again, flints may be hollow, and contain in their cavities either corals or sponges, or masses of that variety of silica known as chalcedony. The latter, it may be mentioned, is a semi-transparent waxy-looking stone, generally with a more or less decided pinkish tinge, and forming mammillated or botryoidal masses. The milky and reddish varieties of chalcedony con-

---

* It also occurs abundantly in the lower part of the Portland stone series of the Isle of Portland, but is there generally less pure, and has the conchoidal fracture less marked.

stitute carnelian, while when it is arranged in differently coloured bands it forms agate. Occasionally small crystals of quartz occur in the hollows of flint.

It has still to be mentioned that in some districts—and more especially near Norwich—in addition to the horizontal layers of nodular flints, there occur in the upper chalk a number of huge cup-shaped masses of flint placed one above another in vertical lines; these masses being locally known as "potstones," and presenting a remarkable resemblance to certain giant sponges called Neptune's cups.

The proportion of the flint to the chalk in the upper chalk of England varies, according to Prof. Prestwich, from four to six per cent. It is important to add that the masses of nodular flint may not unfrequently be found to be traversed by fractures which have subsequently been reunited; thus showing either that the substance must at the time have been in a semi-plastic condition and capable of reunion, or that the fracture has been united by the subsequent deposition of siliceous matter. It must also be mentioned that, as a rule, the white coating is confined to the outer surface of the nodules, and that we do not find layers of pure flint overlain with white coats and then again by other similar layers; thus indicating that the formation of the white coat was the final act in the development of flint.

With these observations on the structure and mode of occurrence of flint, we are in a position to enter on the more difficult subject of its origin. From the very first it was recognized by all geologists that such a peculiarly hard and homogeneous substance as flint, occurring in irregular nodules among the pure white chalk, could not have been deposited in its present condition directly from the waters of the cretaceous sea; and the problem therefore presented itself to account adequately for its mode of formation. The problem soon resolved itself first of all into two questions, namely, whether the silica was originally part and parcel of the chalk as first deposited and that it subsequently gained its present condition, or whether it was added at certain intervals during the deposition of the chalk from a totally different source, or was introduced subsequently to the deposition of the whole.

I believe I am right in saying that there is still a current notion among many persons who are not scientific geologists that flint is a kind of igneous rock ; and its hardness and superficial resemblance to some kinds of obsidian may at first sight lend some countenance to such an idea. Nevertheless, it is almost superfluous to add, such a notion is a totally erroneous one, and the mode in which flint occurs—without altering the limestone with which it is in contact, or the fossils which it contains—is of itself sufficient to refute any such origin.

From the hardness and insolubility of flint, it would appear a very natural inference that silica in all forms is likewise insoluble, but, as proved by siliceous springs, silica in a certain condition is freely soluble in alkaline waters. This naturally suggested to the earlier geologists that flints had been deposited by the aid of hot springs or heated waters in the cretaceous sea at certain intervals during the formation of the chalk sea. Thus, the late Dr Mantell wrote that "the nodules and veins of flints that are so abundant in the upper chalk have probably been produced by the agency of heated waters holding silex in solution, and depositing it when poured into the chalk sea." It will, however, be obvious that there are very serious difficulties in accepting any such explanation. In the first place, we should require the introduction of floods of heated water containing silica at certain irregular intervals during the whole period of the deposition of the upper chalk ; and it would also be essential that these waters should have extended over vast areas of ocean. Secondly, the mode in which nodular flint occurs could not adequately be explained by any theory of this kind ; while it would leave the origin of the tabular flint, which we have seen may occur in joints and fissures, totally unaccounted for. Another idea was that the silica had been, at least partially, introduced from above after the deposition of the chalk ; but, although this explanation may, and perhaps does, partly account for the formation of some of the upper tabular flints, it is quite impossible that it could explain the formation of the numerous layers of nodular flints throughout the body of the chalk.

There is, however, an explanation which will readily account for all the features presented by the chalk-flints, and requires the aid of no foreign factors in the process. This explanation is based on the phenomenon known as segregation. Now segregation is the tendency presented by a small quantity of one substance, when diffused through a much larger quantity of another substance, to collect together in nodules or strings, which generally accumulate either around some fragment of their own nature or some foreign body—especially an organic one— as a nucleus. We have well-known examples of this segregating process in the huge lenticular calcareous masses termed "septaria," found in the London and Kimeridge clays, and also in the iron-nodules of other formations. Premising that soluble silica has a peculiar affinity not only for any kind of silica, but likewise for gelatinous organic substances (both of which were presented by the sponges of the cretaceous seas), it will be obvious that if we can only satisfy ourselves that at the time of its deposition the chalk contained diffused among its substance a sufficient amount of soluble silica, we shall at once be able to account for the formation of its flint. Now, as we pass upwards in the cretaceous system from the lower green-sand to the upper chalk, we find a gradual change from a completely siliceous to a calcareous rock. Moreover, while in the gault, upper greensand, and lower chalk (in which in the south of England there are no flints) there is a large but decreasing amount of soluble silica, varying from 46 per cent. in the upper greensand to 31 per cent. in the chalk-marl, when we reach the upper chalk with flints such soluble silica is reduced to a mere trace. As it is at the base of the white chalk that the sponges attain their greatest development, and as it is also here that flints first commence, the disappearance of the soluble silica may be safely attributed to its segregation by means of the sponges and other bodies, and its conversion into flint. Prof. Prestwich, writing on this subject, observes that in presence of the siliceous spicules of the cretaceous sponges and their gelatinous animal matter, "the soluble or colloidal silica, dispersed through the soft chalk-mud, slowly segregated from out of the surrounding pulpy mass, and gradually replaced part or the whole of the organic matter, as it decayed away. Nor has it stopped there; owing to the affinity of the particles of colloidal silica amongst themselves, the segregation has not ceased with the replacement of the organic body, but has continued so long as any portion of silica remained in the surrounding soft matrix; whence the frequent excess of flint beyond the interior or body of the shells, echinoderms, &c., and whence also the irregular shape arising from this over-growth of the flint nodules." Next to sponges, echinoderms seem to have afforded the most attractive centres of segregation; and while in some cases only their shells have been filled with flint, in other instances we find a mass of these shells cemented together by a nodule of flint.

In some parts of the Continent, and also in Yorkshire, we find that for some reason or other—not improbably a greater development of sponges and a smaller amount of soluble silica—the segregating process has extended down-wards to the lower chalk, where flints are then found; and we suspect that in such cases analysis would also show in these beds a corresponding absence of free soluble silica.*

With regard to the so-called "potstones" of the Norfolk chalk, some of which may be upwards of a yard in height, with a diameter of a foot or so, the only adequate explanation of their formation that has yet been offered is that they represent gigantic cup-like sponges which have grown one upon the top of the other, as they were successively buried in the newly-formed chalk, and that they have been subsequently silicified by the same segregating process.

We conclude, therefore, that the flints of the chalk were originally an integral portion of the rock itself, which was

---

* Since this was written, some observations have been published tending to throw doubt upon the constancy of the relation between the absence of flint and the presence of free soluble silica in chalk.

then a slightly silicated limestone; and that the present purely calcareous character of the chalk is due to the separation of the silica by segregation. We have, however, still to account for the relatively large amount of soluble silica present in the cretaceous rocks, since this is far in excess of what would have been brought down by most rivers of the present day, in the waters of which the amount of this substance is almost infinitesimal. It has been suggested that the unusual supply may have been afforded by the cretaceous rivers being largely fed by siliceous springs; but although this may have been one factor in the case, a more probable theory is that the drainage area supplying the cretaceous sea with sediment was largely composed of decomposed felspathic rocks, in which the amount of this silica would have been amply sufficient to have furnished the quantity present in the chalk.

It is, however, not only with regard to its mode of origin that flint is of more than ordinary interest. Being an excessively hard substance, it is one exceedingly difficult to be worn to powder by the action of water, and the flint-gravels of the valleys of the south of England, as well as the beaches of our southern coasts, and the numerous Tertiary deposits composed of flint pebbles, remain to us as silent witnesses of the vast denudation of the upper chalk which has taken place in this country. Remembering that the proportion of flint to chalk is only from four to six per cent., and also bearing in mind that all the flints in our gravels have been considerably reduced in size by the action of water, we may fairly say that every cubic yard of pure flint gravel represents the removal of at the very least twenty cubic yards of pure chalk; and to this we have to add all the lower chalk which has been denuded without leaving any solid residue. Moreover, when we recollect that this denuding process has been going on ever since the Eocene period, and that our river gravels only represent a small portion of the flints left by the denudation of the chalk during the latter part of this protracted period of time, we may gain some faint conception of how enormous this denudation must have been.

Although the flint-gravels of our rivers afford some estimate, however faint, of the denudation of the chalk during the Pleistocene period, it would be quite incorrect to assume that the flint pebbles forming the beaches of our southern coasts present a record of the amount of denudation which has taken place during the modern period. We have already mentioned that a freshly-broken flint presents a uniform blackish-blue colour throughout its interior, and any flint pebble on the seashore which had been recently derived from its native chalk would, when broken, present a similar appearance. As a matter of fact we shall find, however, that at least ninety per cent. of such pebbles are stained yellow, brown, red, or black internally, and since most of the flint fragments in many of our older gravels are likewise similarly stained, we shall have little hesitation in coming to the conclusion that our modern sea-beaches are largely derived from the breaking up of such old gravel beds, and the subsequent rounding of these irregular fragments of flint into pebbles by the action of the sea. The staining of the flints is of course due to the large amount of ferruginous matter contained in the gravels, and owing to the banded nature of the original flint it frequently gives rise to an agate-like appearance in the pebbles. Many of the pebbles in our beaches are, however, derived from still older sea-beaches, like the one now remaining at the southern extremity of the Isle of Portland, while others, again, owe their origin to the breaking up of the Eocene Woolwich and Reading beds, which are largely composed of flint pebbles. Sometimes, indeed, fragments of these old beds also occur in the river gravels, where blocks of the Hertfordshire conglomerate—the equivalent of the Woolwich and Reading beds—may be met with.

We have thus abundant evidence of the exceeding indestructibility of flint, and how it may go on from one formation to another to tell, when rightly interpreted, the various steps in the denudation of our country.

In addition to being frequently stained internally, the observer will also not fail to notice that all of the flints in our river gravels have acquired a white or yellow porcelaneous external coating, quite different from the interior; and it is believed by Sir John Evans that this white coating has been produced by the removal of a portion of the flint which was still soluble "by the passage of infiltrating water through the body of the flint." That such a process must have been of inconceivable slowness, and must have required countless years for its accomplishment, goes without saying. We have, indeed, some inkling of how extremely slow this process must be, from the circumstance that the fractured surfaces of the flints built into the walls of our very oldest churches show not the slightest change from their pristine condition. When, however, we examine the chipped flint implements of our river gravels and caves, like the one shown in our first illustration, we find their surfaces altered precisely in the same manner as the flint fragments by which they are accompanied. Hence we gain, from a totally independent source, some idea of the immense antiquity of the period when the old palæolithic hunters inhabited the south of England.

Having thus reached the subject of flint implements, we feel tempted to enter into the consideration of some of their different types and the beds in which they occur, but limits of space forbid our wandering into such entrancing paths. Even, however, without entering into this part of the subject, we trust that what we have written will serve to show that a "Flake of Flint," when considered from all points of view, is to the full as interesting as a "Lump of Chalk."

THE END.

www.ingramcontent.com/pod-product-compliance
Lightning Source LLC
Chambersburg PA
CBHW030110030726
47498CB00007B/2328